NATURAL HOME HEATING

NATURAL HOME HEATING

The Complete Guide to Renewable Energy Options

GREG PAHL

Chelsea Green Publishing Company
White River Junction, Vermont

Designed by Peter Holm, Sterling Hill Productions

Printed in the United States at Vicks Lithograph & Printing Corporation
First printing, September, 2003

10 9 8 7 6 5 4 3 2 1

Printed on Glatfelter's Authors, an 85% recycled paper containing 30% post consumer waste

Library of Congress Cataloging-in-Publication Data

Pahl, Greg.
Natural home heating : the complete guide to renewable energy options / Greg Pahl.
p. cm.
Includes bibliographical references and index.
ISBN 1-931498-22-9
1. Dwellings—Heating and ventilation. 2. Renewable energy sources.
I. Title.
TH7684.D9P34 2003
697—dc22

2003015397

Chelsea Green Publishing Company
Post Office Box 428
White River Junction, VT 05001
Editorial and sales offices: (802) 295-6300
To place an order call: (800) 639-4099
www.chelseagreen.com

For my wife, Joy

"Never doubt that a small group of thoughtful, committed citizens can change the world; indeed, it's the only thing that ever has."

—ATTRIBUTED TO MARGARET MEAD

CONTENTS

ACKNOWLEDGMENTS

While I worked on this project, I met and spoke with some wonderful, enthusiastic people who are committed to helping this country free itself from its dependency on fossil fuels.

I would like to acknowledge the many people who were so generous with their time and advice. This book would not have been possible without people like Max Pellerin, operations manager of Vermont Pellet Stoves, Inc. in Colchester, Vermont; John Van de Vaarst, deputy area director at the Beltsville Agricultural Research Center in Beltsville, Maryland; Leigh Seddon, president of Solar Works, Inc. in Montpelier, Vermont; Clay Turnbull, sales manager of Solar Works, Inc. in Townshend, Vermont; John Hurley, Dog River Alternative Fuels in Berlin, Vermont; Keith Ciampa, executive vice-president of World Energy in Chelsea, Massachusetts; Roy Truesdale, director of operations of the National Biodiesel Board in Jefferson City, Missouri; Craig Issod, webmaster at HearthNet; John Crouch, director of public affairs, and Leslie Wheeler, both of the Hearth, Patio and Barbecue Association; Randy Machia, Sr., Russ Beamish, and Roy L'Esperance of The Chimney Sweep Fireplace Shop in Shelburne, Vermont; Ken Rajesky, marketing manager of Hearthlink International in Randolph, Vermont; Paul Butler of Vermont Alternative Energy Corporation; Lloyd Nichols of Tarm USA, Inc. in Lyme, New Hampshire; Beverly J. Marois, administrator of The Masonry Heater Association of North America in Randolph, Vermont; Norbert Senf of Masonry Stove Builders in Shawville, Quebec, Canada; Jerry Frisch of Lopez Quarries Masonry Heaters/Firecrest Fireplace Corp. in Everett, Washington; Leslie Wheeler, executive director of the Pellet Fuels Institute in Arlington, Virginia; Steve Walker, president of New England Wood Pellets in Jaffrey, New Hampshire; Bruce Brigden of Pinnacle Stove Sales in Quesnel, British Columbia, Canada; Mike Haevner, president of American Energy Systems, Inc. in Hutchinson, Minnesota; Sara Quinn, communications director, and Conn Abnee, executive director, both of the Geothermal Heat Pump Consortium, Inc. in Washington, D.C.; and Ed Dooley, vice president of communications and education for the Air-Conditioning and Refrigeration Institute (ARI) in Arlington, Virginia.

I also want to thank professor Richard Wolfson and assistant professor

Helen Young from Middlebury College for their assistance, as well as Jim Barbour of Shelburne, Vermont; Terry Mason of North Wolcott, Vermont; David Lyle of Ackworth, New Hampshire; Stuart Davies of Ashland, Oregon; and Rick Kerschner, director of preservation and conservation for the Shelburne Museum in Shelburne, Vermont, for their comments and observations.

I would also like to offer my sincere thanks to all the wonderful folks at Chelsea Green Publishing who helped to bring this book to completion. In particular I want to thank Alan Berolzheimer, Jim Schley, and Fern Marshall Bradley, my editors, who helped guide me through the process. Various parts of the excellent but long-out-of-print work, *The Book of Heat,* by William Busha and Stephen Morris, published in 1982 by the Stephen Green Press, were excerpted or reworked for use in chapters 8 through 11; I thank Stephen Morris for his kind permission to make use of this valuable resource.

I also want to thank the many, many people who helped supply the illustrations, and anyone else I may have forgotten to mention. All of your contributions, both large and small, are greatly appreciated.

Last, but by no means least, I want to thank my wife, Joy, for chasing down obscure facts, proofreading, making suggestions, and generally putting up with me while I was trying to meet deadlines.

PREFACE

I've been interested in renewable energy for almost 30 years. Back in the mid-1970s, I built my first off-the-electric-grid home in Monkton, Vermont. I pumped water by hand from a dug well, used kerosene lamps for lighting, and heated with wood. In 1979, I moved to a home in the middle of the woods at the end of a half-mile-long driveway in the remote and beautiful mountain town of Lincoln, Vermont. The commercial power lines were too far away, so after investigating my options, I finally decided to install a wind turbine atop an 80-foot guyed steel tower to provide for my modest electrical needs. This home had a wonderful spring-fed water system and was heated exclusively by wood. Subsequently, I have lived in homes that were grid connected, but generally used wood as a primary or secondary source of heat. And I've never lost my interest in renewable energy.

In November 2000, my wife, Joy, and I bought a house in Weybridge, Vermont. We chose this home primarily for its quiet, edge-of-town location on a dead-end street, and *not* for its heating systems. This modest Cape-style house has an attractive, but not very functional, fireplace. Although the fireplace works great as a smoke-alarm tester, trying to figure out how to use it as a heating appliance has been a challenge. But that's not all. Our house also features its original 1956 American Standard oil-fired boiler still chugging away in the basement. The first time our local heating service technician came over to give the boiler its annual cleaning and inspection he said, "Is that thing *still* here?" I knew we had a problem. The technician suggested that a new high-efficiency boiler would be the best approach in the long run but admitted that a few repairs would probably keep the unit going for a few more years. Actually, the old veteran was still working reasonably well, but needed a new oil burner, as well as some new electrical controls, in order to bring it up to code. Being a bit short of cash, we decided on the temporary fix to give ourselves more time to see what our other options might be. Thus began our search for a replacement heating system that would not be reliant on imported oil—or on any other fossil fuel.

The following summer, our interest in sustainable energy led us to attend SolarFest 2001, an annual renewable energy fair held in Middletown

Springs, Vermont. It was at SolarFest that I first learned about the possibility of burning biodiesel fuel in an oil-fired boiler or furnace for home heating. I was intrigued. It was also at SolarFest that I first met Stephen Morris, the publisher and president at Chelsea Green Publishing. For some time, I had been searching for a new publisher that specialized in environmental titles, and Chelsea Green looked like it might be a good match. It was. Our first brief conversation at SolarFest was followed by a series of e-mails and eventually a lengthy meeting in White River Junction, where the idea for this book was born. I combined my existing interest in finding a new renewable heating system for my home with Chelsea Green's interest in a book on the same subject. Sometimes things just seem to fall into place.

As I researched this book, my knowledge of renewable home heating strategies expanded well beyond woodstoves to include a wide range of options, some of which I had been only dimly aware of. My research also confirmed what I already knew, which is that renewables are definitely available now. And renewable fuels work just fine, even for home heating. The main obstacles to their widespread use are no longer technical; they're mainly political. In order to deal with that problem, there needs to be a major educational initiative to supply people with the information they need to make intelligent energy-related decisions. I hope that *Natural Home Heating: The Complete Guide to Renewable Energy Options* will play a role in this important initiative. It's been an exciting adventure for me, and I hope that after reading this book, you will share my enthusiasm for renewables—and for Chelsea Green.

INTRODUCTION

The oil embargo of 1973 sent shock waves through the United States economy and dramatically inflated energy prices nationwide. While millions of Americans sat in long lines at gas stations waiting for their ration of gasoline, millions of others searched for less expensive, alternative ways to heat their homes during the winter. Many of those people eventually ended up returning to the use of wood or coal in order to stay warm during the winter months, sparking a renaissance in wood and coal stoves as well as furnaces. The search for alternative energy sources also spurred the development of wind-, hydro-, and solar-powered technologies nationwide. For a brief time, it appeared that this country was on its way to making the inevitable shift from nonrenewable fossil fuels to sustainable energy.

But when the embargo ended, oil prices plummeted, and the Reagan Administration pulled the plug on the alternative-energy sector by eliminating the energy tax credits and incentives that had helped to encourage its growth. The budding alternative-energy industry collapsed. Many people continued to burn wood and coal for home heating, but the growing concerns about stove-related air pollution, as well as frustration with the extra labor and household mess, gradually eroded support for these fuels. The return to cheap oil was hard for most people to resist.

Then, beginning in 2000, the California electricity deregulation crisis, along with a sudden increase in the price of natural gas, brought energy issues back to center stage. The sticker shock of the 2000–2001 heating season was still fresh in the minds of most Americans when the tragic events of September 11, 2001, unfolded on television screens across a shocked and horrified nation. As the twin towers of the World Trade Center collapsed in clouds of smoke and debris, a false assumption that most Americans had long maintained—that life in the United States was somehow separate from the rest of the troubled world—crumbled as well. The attacks, and the subsequent U.S. military campaigns in Afghanistan and Iraq, also served as urgent reminders about the dangers of continued reliance on oil, especially imported Middle Eastern oil. It was déjà vu all over again.

Despite the terrible tragedy, the events of September 11, along with the California energy crisis, have had the effect of refocusing national attention on our overreliance on fossil fuels. It has become increasingly clear to a growing number of people that expanding the supply of domestic coal, oil, and natural gas to meet demand only prolongs our dependence on finite, nonrenewable resources. This strategy simply doesn't make any long-term sense at all. But how can we break our addiction to fossil fuels? The obvious answer—once again—is renewable energy.

What's more, since September 11, many people no longer view a shift to renewable energy sources as simply an environmental issue. In addition to making long-term business sense, support for renewables has now been elevated to the status of a national security issue. It's now rightly viewed as being patriotic to support the development and widespread adoption of a broad range of sustainable energy initiatives that will ensure a steady, safe domestic supply of energy well into the future.

The revived interest in sustainable energy issues has also generated increased interest in renewable home heating. That's a good thing, because it is estimated that one-fourth of all the energy used in the United States is consumed in heating buildings. Billions of Btus are expended every heating season in this country. Unfortunately, most of this energy still comes from nonrenewable resources such as coal, oil, and natural gas. In addition, as we have seen in the past few years, the prices of these fuels are becoming more and more unstable—and the supplies increasingly uncertain. Worse yet, these fossil fuels are now almost universally recognized as major contributors to global warming and a wide range of health concerns. Heating systems in the United States emit a billion tons of carbon dioxide and about 12 percent of the total sulfur dioxide and nitrogen oxides emitted in the nation. Although it's still hard for many Americans to understand or accept this fact, the end of the fossil-fuel economy is not that far off. We need to face this reality squarely and start to make the transition to renewables sooner rather than later, before the fossil-fuel energy market descends into total price chaos. The California deregulation crisis gives us just a hint of what that would be like. It's not a pretty picture. That sort of scenario on a national scale is something we absolutely must avoid.

This is an especially vital concern when it comes to home heating. In most of the United States—especially the northern part—heating your home can literally be a matter of life and death. Heating is the largest energy expense in most homes, accounting for almost two-thirds of annual energy bills in colder parts of the country. But how can homeowners reduce or eliminate the use of fossil fuels in heating their homes? Although there is no single, easy answer to that question, happily there are many interesting and viable alternatives. That's what this book is all about. *Natural Home Heating: The Complete Guide to Renewable Energy Options* is a comprehensive consumer's guide to home heating systems that rely on renewable sources of energy.

Among renewable sources, the most inexhaustible and cheapest is also the oldest—the sun. And the best way to take advantage of the sun's heating capacity is to live in a carefully designed and constructed solar home. Unfortunately, the sun doesn't shine very often in some parts of the nation. And in other, mostly northern regions, even a well-designed passive solar home will almost certainly need some additional source of heat during the dark, cloudy, frigid days of winter. But there's another problem with the solar solution. Most people (myself included) live in homes that, due to their basic design, take little or no advantage of solar heating. What about us? Are we doomed to rely on fossil fuels until the supplies run out, or until we can't afford to pay for them any longer? Fortunately, that's not necessary.

Natural Home Heating covers all of the options for renewable home heating. If you're thinking about building a new home, you'll find that there is a wide range of renewable systems that will keep you comfortably cozy on those long winter nights. If you live in a home with less-than-ideal solar design or exposure, you'll appreciate the information on retrofit strategies that the book offers. And regardless of the type of house you live in, if your furnace or boiler is wheezing through its final days, you can find information in *Natural Home Heating* about how to choose the best renewable replacement system. For those who live in a mild climate, this book offers some excellent ideas on how to meet both cooling and heating needs with renewable systems. I'll also offer some energy conservation strategies that will cut your heating bills and increase your comfort at the same time. Every little bit helps. And everyone has a role to play in the national shift to renewables.

We'll begin this exploration in part 1 with basic information about what makes for a comfortable home. We'll look at heat transfer, temperature, humidity, and air movement. Then, I'll introduce you to the main components of both nonmechanical and mechanical heating systems and explain what issues you need to keep in mind while trying to choose a system. We'll also take a comparative look at the potential heat sources for a home heating system, such as stoves, furnaces, boilers, masonry heaters, heat pumps, and so on. In addition, we'll check out the fuels that these devices can utilize and their relative advantages and disadvantages. You may be surprised by some of the many options that are available.

In part 2, we'll examine the key renewable heating source: the sun. I'll explain the fundamentals of passive solar heating design, including orientation, absorbers, heat sinks, distribution, and controls. We'll also look at the basic components that make up active solar heating systems and how they can be combined into a variety of home heating systems. I'll also explain which strategies are most appropriate for new construction and which are better suited for renovations or retrofits.

In part 3, we'll take a look at heating your home with wood. I'll cover everything you need to know about obtaining, handling, storing, and burning wood. Then, because it's so important from a safety standpoint, we'll look at stove placement, clearances, hearthpads, chimneys, stovepipe, and more. Once we've covered the basics of heating with wood, we'll check out the vast array of stoves, furnaces, boilers, combination systems, fireplaces, and masonry heaters from which you can choose, with special attention to costs and comparative advantages and disadvantages.

In part 4, we'll examine the exciting possibilities of using other forms of biomass (beyond cordwood) as a renewable home heating strategy. I'll explain the various fuels currently available, including wood pellets, corn, and other grains. Then, we'll explore the rapidly expanding world of pellet stoves, pellet fireplace inserts, pellet furnaces, and pellet boilers. We'll compare prices and examine the advantages and disadvantages of these systems.

Finally, in part 5, we'll take a look at geothermal home heating and cooling. I'll explain the fundamentals of geothermal energy and heat pumps. We'll examine air-source and ground-source heat pump systems. I'll tell you how to choose the correct heat pump for your geographic location and how to size it and locate it properly. And we'll also take a careful look at initial installation expenses versus long-term operating costs as well as the advantages and disadvantages of these systems.

Throughout the book, I have included conversations with people who are well versed in renewable home heating strategies. These movers and shakers in the industry will share their comments, observations, and knowledge with you. Along the way, I'll also offer installation, operation, maintenance, and safety tips that will help you to avoid common mistakes that others have experienced with their renewable heating systems. Although this is not a technical manual, *Natural Home Heating* will give you the basic information that you need to make intelligent, informed decisions about your renewable home heating options. And that's an important point. Many people ask, "What's the best heating system for *my* house?" The answer, usually, is that you have more than one option. You will need to decide which one makes the most sense for you and your particular situation. To assist you in the process, I've included some extra tools at the end of the book. To help you understand system details, a thorough description of home heating terms can be found in the glossary. For those who wish to learn more, check out the bibliography for an extensive listing of other books, and the comprehensive listing of additional sources of information included in the "Organizations and On-Line Resources" section.

I hope that by the time you finish reading this book, you will understand that it is possible to switch to renewables and stay comfortably warm, while also saving money and the environment. Living an environmentally sensitive lifestyle does not require you to be uncomfortable. In fact, you'll find that heating your home with renewables can be a fascinating and enjoyable adventure—as it has been for me. Making this change does require that you become more aware of your choices and that you make informed decisions based on that knowledge. Knowledge is power. And some of that power can keep you quite comfortable in January. So, make yourself a cup of hot tea or hot chocolate (or your favorite beverage), sit down on the sofa, pick up a good book (preferably this one), and settle in for a long winter's evening. And if you're a little chilly, throw another chunk of wood in the stove, or perhaps just turn up the thermostat on your pellet-fired boiler. You'll have the satisfaction of knowing that you are doing your part to help create a more secure future for everyone.

PART ONE

HEATING BASICS

CHAPTER 1

Heating 101

It all started with fire. Possibly as early as 1.4 million years ago, our distant ancestors in Africa may have used fire to cook meat, harden wooden spear points, and protect themselves from hungry predators. But fine-tuning the use of fire took a long time. Almost a million years passed before *Homo erectus* began to use fire for heating. Home heating in that era consisted of a wood fire on the floor of a cave.

Humans eventually moved out of caves and into crude houses heated by a fire in the middle of the floor. Smoke escaped through a hole in the roof. It was not much better than the caves, but at least there were fewer bats flying around.

During the Roman era, earthenware stoves were the heat source in homes in central and northern Europe. Although they were definitely an improvement over open fires, these stoves still tended to rely on the hole-in-the-roof strategy for smoke removal. The Chinese developed the first cast-iron stoves around 200 A.D., but this technology didn't catch on in Europe until the 15th century. By the 1700s, stove design had advanced significantly, resulting in a wide variety of improved models of both masonry and cast-iron stoves.

A crude, smoke-filled home was (marginally) better than living in a cave.

In the 1800s, just when wood burning reached the high point of its development, it was eclipsed by a fuel switch to coal. At the time, people viewed the use of this newly discovered fossil fuel as a great technological advance that also solved the problem of dwindling firewood supplies in many parts of Europe. Although some people—especially in rural parts of the United States—continued to use wood, coal gained popularity and helped to propel the gradual shift to central heating, especially after the development of steam heating around 1850. Central-heating technologies based on coal became increasingly sophisticated and effective. Then, along came oil, gas, and electricity. By the 1950s, most of the developed nations (the United States in particular) depended almost completely on "modern" fossil fuels for heating as well as for transportation and generating electricity. In hindsight, this total dependence on fossil fuels doesn't seem like such a great idea.

This book is about making the transition from fossil fuels to renewables. In order to understand the relative advantages and disadvantages of heating systems that use renewable fuels, and to choose the one that's right for your home, you first need to understand how heating systems operate.

It's All about Comfort

The purpose of having a home heating system is personal comfort. Admittedly, most early heating strategies didn't rate very high in that department. Whether in a cave or an early home, an open fire did not provide ideal conditions for regulating body heat, which is a key element in human comfort. It's important to understand that our bodies are like portable heaters. We burn (digest) food to provide the energy that keeps us alive. Digestion and other metabolic processes produce heat, and that heat has to be dissipated into the surrounding environment in order for us to feel comfortable. So, ironically, it's really the regulated *loss* of body heat that determines how warm or cold we feel, rather than the other way around.

Our comfort level tends to vary from person to person. Variables that influence comfort include such things as activity, clothing, humidity, air movement, air temperature, and the temperature of surrounding surfaces. Infants and older people tend to be less tolerant of extremes in temperature. Teenagers, with their high metabolic rates (to say nothing about hormones), can be quite comfortable in cooler temperatures. It's not unusual for one member of your family to be happy with the thermostat set at 68 degrees Fahrenheit, while another prefers a setting of 72 degrees Fahrenheit. Sound familiar?

A change in any of the variables that affect our comfort level changes the relationships of the others. That's why simply maintaining a particular temperature in a room does not always ensure comfort. When you understand the subtle factors that influence comfort, you will have an easier time appreciating the qualities of various heating systems.

Heat Transfer

When people think about heat, fire is one of the first images that come to mind. In one form or another, fire is central to most heating strategies. In order to understand how the heat from fire warms your home, you need to understand the concept of **heat transfer.** Here's a key point to remember: Heat naturally flows from a warmer area (or substance) to a cooler area (or substance). This transfer of heat tends to lower the temperature of the warmer area/substance and raise the temperature of the cooler area/substance. The flow of heat can be up, down, or even sideways, but the direction of flow is always from hotter to cooler.

In order to further understand heat transfer, let's look at the three ways heat transfer takes place: conduction, convection, and radiation.

Conduction

Conduction is the flow of heat through solids, liquids, or gases—always from hot to cold. When two objects touch, heat moves from the warmer to the cooler object. For example, if you stick a metal poker in a fire, the heat from the fire will flow along the poker and eventually to your hand. The ability of the poker to transfer that heat is called conductivity. Some substances, such as metals and water, are good conductors; others, such as wood and air, are not. Conduction does not play a large role in home heating systems.

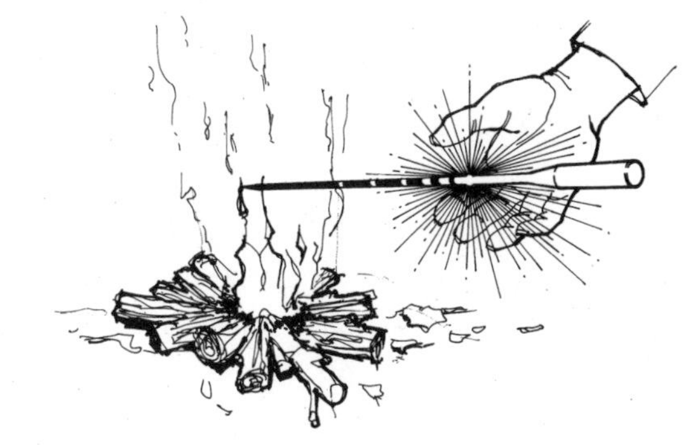

An example of conduction.

Convection

Convection is the movement of heat due to motion in liquids (such as water) and in gases (such as air) from warmer to cooler areas. When molecules of a fluid or gas are heated, the heated molecules expand. The heated portion of the fluid or gas weighs less than the surrounding, unheated portion, and thus the heated portion rises. Unheated molecules replace the space that was filled by the heated molecules. The heated liquid or gas may move naturally or be forced to move by pumps or fans. Convection currents play an

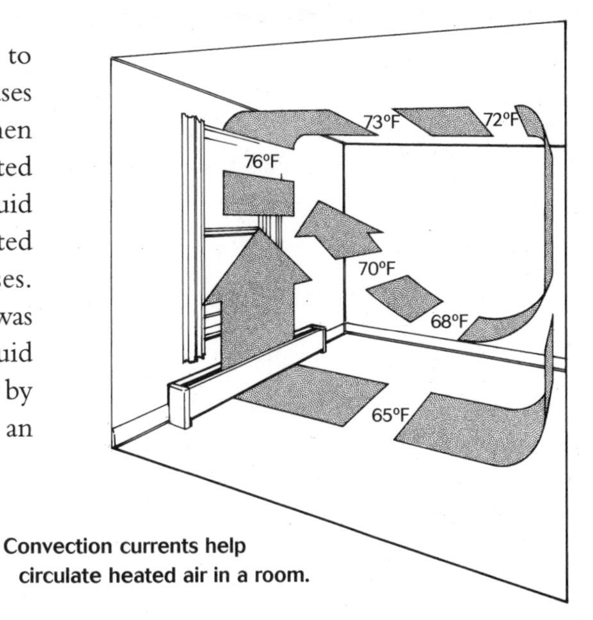

Convection currents help circulate heated air in a room.

important part in home heating systems, particularly in baseboard heating elements, and also in the draft of a chimney.

Radiation

Radiation is the transfer of heat by electromagnetic waves. Radiation is different from conduction and convection in that the substances that exchange heat do not need to touch each other. Radiant heat transfer can take place through air, through a vacuum (such as outer space), or even through a solid material such as glass. As long as two objects that are at different temperatures can "see" each other, they will tend toward the same temperature through radiation. Radiation does not heat the air through which it passes, but it does heat any objects that it strikes, such as your body, walls, floors, and home furnishings. You feel radiant heat from direct sunlight; from an open fire; or from radiant floor, wall, or ceiling heaters. Radiation plays a major role in a variety of home heating strategies.

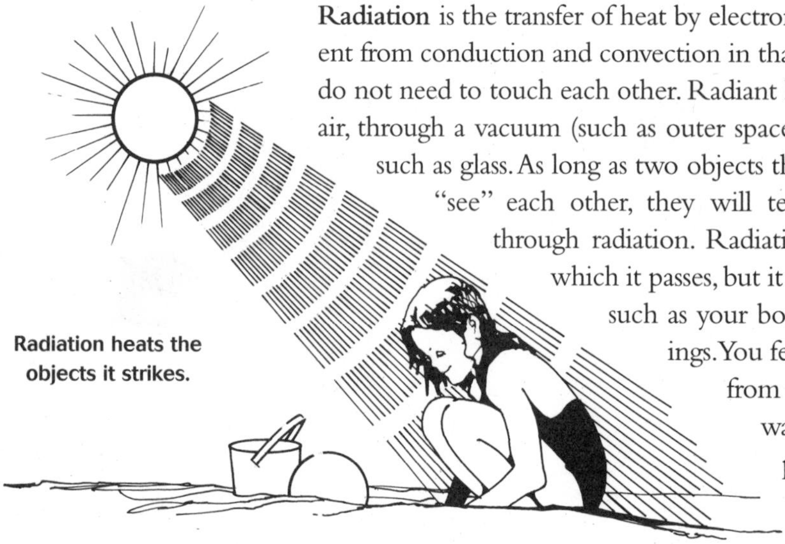

Radiation heats the objects it strikes.

Temperature

Simply stated, temperature is a measure of heat intensity. In most homes, the temperature in any room is uneven. There is often a significant difference between the air temperature at floor level and at ceiling level; this difference is known as **temperature stratification.** The air temperature needed for comfort depends on several variables, including the clothing you are wearing, your physical activity, your age, and the amount of radiation to or from various room surfaces. During the winter, you might be comfortable in a room with an air temperature of 70 degrees Fahrenheit, yet if that room contains large glass windows, you might need a higher air temperature in order to feel comfortable. That's because glass allows radiant heat transfer: your body is actually radiating heat to the outdoors through the window glass, giving you a cooling sensation.

A related measure of temperature called **mean radiant temperature** is one of the most important elements that affect your comfort. Mean radiant temperature is the average temperature of the surfaces that surround you in your home, such as the walls, floor, ceiling, and windows. (Windows are usually the coldest surfaces in a room.) Mean radiant temperature is also a measure of the radiant heat exchange between you and your surroundings, and it can vary from one place in a room to another.

Humidity

Humidity refers to the moisture content of the atmosphere. **Relative humidity** is a measure of the amount of water vapor in the air compared with the amount of water vapor that the air can hold when completely saturated. The maximum amount of water vapor that air will hold depends on temperature. The higher the air temperature, the higher the amount of moisture that the air will hold. The optimum relative humidity range for humans is 30 to 60 percent. Relative humidity above 60 percent can lead to discomfort, especially during hot summer weather.

Humidity levels play a key role in determining comfort levels in home heating (and cooling). In the winter, high humidity levels in your home can encourage molds and mites that pose a health hazard. At the other end of the spectrum, low relative humidity (especially during the winter) can cause your body to dehydrate, resulting in dry mucous membranes and dry skin. When humidity levels in your home are too low, higher air temperatures are needed to compensate. Under these circumstances, adding humidity (with a humidifier or other source of moisture) is a good idea.

Air Movement

The movement of air generates heat loss and produces a cooling sensation on your skin by causing evaporation of moisture from your skin into the air. The rate of this evaporation (and heat loss) is related to the rate of air movement and the relative humidity. The movement of cool air down your neck and around your ankles plays a crucial role in how comfortable you feel in your home during the winter, because these parts of your body are especially susceptible to drafts. During the winter, too much air movement can cause discomfort. This can be a problem with poorly designed forced hot-air heating systems.

Energy Units

Units of measurement for heat energy can be a confusing topic. One problem is the many different units of measurement. Another problem is that heating fuels are sold on the basis of units of measure that often have no relationship to a unit of heat energy. Natural gas, for example, can be sold by volume (cubic feet) or by energy units, while oil is sold by the gallon, propane by weight or volume, electricity by the kilowatt-hour, wood by the cord, and wood pellets by the pound or ton. However, as you'll see in chapter 4, it is possible to use energy units to compare heating fuels.

The **British thermal unit (Btu)** is the basic energy unit in the Imperial system. A Btu is the amount of heat required to raise the temperature of 1 pound of water by 1 degree Fahrenheit. (A related unit of measure, the therm, equals 100,000 Btu.) In the United States, most heating devices are rated in Btu per hour (Btu/hr), the maximum amount of energy that the unit will produce in one hour (sometimes listed as MBH, or thousands of Btu per hour).

In the metric system, the **joule** is the basic energy unit. The joule is a very small unit; one joule equals only 0.00095 Btu. One thousand joules, called a kilojoule, is slightly smaller (by 5 percent) than 1 Btu. One billion joules is known as a gigajoule (GJ). We won't be discussing joules much, but it's helpful to know what joules are because they're often used in descriptions of heating appliances manufactured in Europe.

The basic unit of measure for electricity is the kilowatt-hour (kWh). One kilowatt-hour is one kilowatt of electric power delivered for one hour. Or, put another way, ten 100-watt lightbulbs burning for one hour consume 1,000 watt-hours or one kWh. One kWh equals 3,413 Btu or 3,600 kilojoules.

Heat Loss

Before you begin reading about heating system specifics, you'll want to determine how much heat your home loses. If your home didn't lose heat, you might not even need a heating system (or at least not a very big one), so heat loss is a key issue. In fact, I can't overemphasize the importance of striving for a well-designed and well-insulated house that is suited to its climate and environment. This kind of house loses very little heat. Unfortunately, most of us don't live in optimally designed and constructed houses, so we have to deal with heat loss—sometimes quite a lot of it. This heat loss occurs through the various elements of a home's shell (or envelope) and through air leakage or infiltration.

During the winter, heat inside your home is constantly trying to escape. This lost heat needs to be replaced by your heating system. How large should that system be? And how do you know how much heat your home loses? One way is to estimate by using a rule of thumb suggested by your contractor, but I don't recommend this. The rule of thumb might be correct; then again, it might be wrong. One of the most common (and serious) mistakes in home heating is installing an oversize system—often the result of an incorrect rule of thumb being applied.

A much better way to determine heat loss is to apply a formula for heat loss calculation:

$$HL = (A/R) \times dT$$

HL = heat loss (expressed in Btu)
A = surface area of component
R = thermal resistance value
dT = temperature difference

Frankly, my eyes tend to glaze over when I work with mathematical formulas, so I'll try to explain this as simply as possible.

The **design-day heat loss** of your house is the total amount of heat lost on a "design day" for your location (generally) assuming an indoor temperature of 72 degrees Fahrenheit. The heat loss formula above takes into account such issues as the surface area of various components of your home (A), resistance to heat flow (R), and the temperature difference between the interior and exterior (dT). The temperature difference is based on long-term (often 30-year) weather records.

The heat loss calculation is computed for each home component that is in contact with the exterior, such as walls, windows, floors, ceilings, doors, and basement surfaces. The different insulating levels of these components are factored in as well (as resistance to heat flow). Many heating contractors and heating-appliance dealers offer heat loss calculation services. Or, you can buy one of several excellent software programs that calculate heat loss.

The Thermostat Myth

There is a widely accepted bit of thermostatic folklore that I want to lay to rest once and for all. Turning up a thermostat to 90 degrees Fahrenheit or higher will not cause your house to warm up any faster than setting the thermostat at 70 degrees Fahrenheit (or whatever your normal comfort level may be). A thermostat is generally a simple on/off switch. Your heating system cranks out heat at the same steady rate until it reaches the temperature set on the thermostat, no matter what that setting is. So if your house is chilly, just set the thermostat for the desired temperature—and relax.

In an older house, 30 to 50 percent of total heat loss may take place through the windows.

Heat Gain

If your house is losing heat like a sieve, your ideal strategy would be to reduce that loss by installing more insulation and better windows and by sealing up drafty cracks and crevices. Adding insulation to your home is relatively inexpensive and can cut your heating bills substantially. A few hundred dollars' worth of insulation can often pay for itself in just a few years. Many utilities offer free or low-cost energy audits for residential customers in most states. In the long run, it's a very cost-effective strategy, and I recommend it highly.

Assuming you've already reduced heat loss (or that your home is newly built), you need to figure out how to replace the heat your house is losing. This is where **design-day heat load** comes in. You determine design-day heat load by adding together the heat loss figures for all components of your house. The resulting figure (in Btu/hr) tells you how much heat your heating system needs to provide at this design temperature. From that, you can determine the size of the heating system your house will require, since heating appliances are normally rated in thousands of Btu/hr. Keep the size of the system you need in mind as you consider your choice of heating system.

Nonmechanical Heating

In heating your home, simple is often better, and the use of the sun for heat can be quite simple. Passive solar heating is the primary and simplest nonmechanical heating strategy available. However, a properly designed home can also use a variety of other nonmechanical heating strategies, such as woodstoves. The main elements of a nonmechanical heating system are a heat source, a component that stores heat, and a means of distributing heat.

Source

In a passive solar house, the primary heat source is the sun. Even in a nonsolar-designed home, the sun can add some heat through south-facing windows.

Simple wood-burning stoves, as well as fireplaces and masonry heaters, can also be low-tech nonmechanical sources of heat in both solar and nonsolar homes. (In this context, I'm talking about a woodstove that does not rely on any electrical or other mechanical components.)

Storage

One of the key design elements in most nonmechanical heating systems is a means of storing heat that has been collected or generated, so that it can be released later (usually at night). Heat storage is normally accomplished by warming a **heat sink,** which can be water, brick, concrete, adobe, rocks, or other materials (see chapter 5 for details on heat sinks).

Distribution

The warmth stored in a heat sink needs to be distributed back into the living space. In nonmechanical systems, this is accomplished either by radiation or by natural convection currents. Nonmechanical heating systems need to be carefully designed (ideally, along with the house) in order to work effectively.

Mechanical Heating

You will discover that the range of choices for mechanical home heating systems can boggle the mind, but almost all of these systems have three elements in common: a heat source, a means of distributing heat, and a mechanism for controlling heat distribution (some mechanical systems also store heat). All of these elements need to be properly designed to work together, otherwise the system will not do its job efficiently.

Source

At the heart of every mechanical heating system is a source of heat. In an active solar (usually mechanical) heating system, the source is the sun. In most other systems, the heat source can be a stove, furnace, boiler, water heater, or heat pump—or possibly a combination. Often referred to in the trade as a "heating appliance," these heat-generating devices frequently are capable of burning or using more than one type of fuel. The choice of appliance is often determined by what type of fuel is locally available.

Storage

Mechanical systems usually accomplish heat storage by pumping a hot liquid or blowing hot air into a heat sink, where it remains until it is needed (usually at night).

Distribution

Whether or not your heating system relies on heat storage, the heat from the source needs to reach the living spaces in your home. In a central hot-water heating system, pipes carry the hot water to radiators or baseboard heaters, where the heat is then distributed to the room by radiation or convection (or both). In forced hot-air systems, the hot air travels through ducts and the heat is distributed by convection.

Controls

Every heating system needs some sort of control over the amount of heat it provides; otherwise you'd end up alternately freezing and roasting. Simple **thermostats** are temperature-operated switches that turn your heating system on or off in response to temperature changes in your home. Most modern homes use some type of thermostat for heating (and sometimes cooling) control. Complex electronic thermostats are programmable, with multiple heat-zone controls.

To be effective, a thermostat should be located on an inside wall about 5 feet above the floor and away from drafts, heat-producing appliances, and direct sunshine. Ideally, it should also be located in the part of your home where you tend to spend most of your time; for example, the living room. Hint: Even a well-designed heating system may not run well if the thermostat is poorly designed, so don't try to save a few dollars by buying a cheap thermostat.

CHAPTER 2

Types of Heating Systems

Trying to pick the best heating system for your home can be a real puzzle. If you believe the advertising claims made by most manufacturers, *every* heating system on the market is the best. Of course, the real issue is not which system is the best; rather, it's which heating strategy makes the most sense for *you*. As you learn about heating systems, you may find that any of several different systems—or perhaps a combination of systems—could meet your needs. You will also discover that some systems or strategies make more sense for new construction, while others are better suited for retrofits or replacement situations.

Choosing a System

If you are building a new home, you have the advantage of starting from scratch to design the most cost-effective renewable heating system that will match your energy-efficient home design. If you are renovating or replacing an old system, however, you will have to work within an existing building that may impose some structural or other restrictions.

It's important to understand that your building contractor (or even your heating contractor) may not be knowledgeable about all types of heating systems. Contractors tend to recommend and install systems with which they are familiar. That may (or may not) include the system you have in mind. I'm not suggesting that you ignore your contractor's advice; it's probably sound. But you may have to be proactive in order to find the right person to answer questions about systems that aren't popular in your area. Of course, there may be good reasons why those systems aren't popular where you live, but then again, there may not. It's important to explore all your options before you make a final decision. If you are committed to an environmentally responsible lifestyle, that may be another key factor in your decision-making process, regardless of what your contractor may advise.

Comfort and Lifestyle

As you will recall from chapter 1, the primary goal for any heating system is comfort. Consider lifestyle issues that relate to comfort as you try to figure out what heating system will be best for you and your family. For example, what sort of clothes do you normally wear indoors during the winter? Do you tend to wander about barefoot in light clothing, or do you prefer flannel or woolens? Do you have very young or very old members in your family who tend to be more temperature sensitive? Do you spend time sitting on the floor with children or pets? All of these issues may have an effect on your choice of heating system components. For example, if you or your children play on the floor often, you may want to consider a radiant floor system.

House Design Issues

Design factors such as amount of insulation, floor materials, and type and size of windows or sliding glass doors also affect your comfort. Large expanses of glass, in particular, will put more demands on a heating system, because glass is always the coldest surface in a home during the winter. Sizable windows substantially reduce the mean radiant temperature in a room, making it more difficult to maintain a comfortable environment. Installing high-performance windows or additional heat sources along the exterior wall(s) below large windows may be necessary.

It's also hard to maintain a comfortable temperature at floor level in a room with a high ceiling, due to temperature stratification. Combine high ceilings with post-and-beam construction, and you create a particular design challenge for some heating systems, especially those that utilize ductwork. I realize that cathedral ceilings are popular, but I don't recommend them, at least not in cold northern climates.

The size of your home is another design factor that will influence your choice of a heating system. A huge, three-story home can be very difficult to heat unless it's carefully designed. A small, one-story home built on a slab is usually easier and less expensive to heat. Also, it may be better suited to certain types of heating strategies, such as radiant floors or stoves. Most modern homes that are relatively draft free and well insulated and that have an open-living-space design may not need a traditional central-heating system. A stove or masonry heater, for example, might serve nicely as the primary source of heat.

Costs

"How much is it going to cost?" is one of the most common (and obvious) questions you'll ask about a heating system. The answer, however, is not so obvious. That's because there are two important factors to weigh: capital costs and operating costs. Generally speaking, the more you spend up front on capital costs, the less you will spend later on operating costs. For most heating systems, your operating costs will far exceed your original capital costs. When choosing a heating system, most people are tempted to look only at capital costs and ignore operating costs. However, we'll look closely at both types of costs.

Capital Costs

Capital costs are the costs of purchasing and installing a heating system. Determining the price of a furnace, for example, is straightforward, yet trying to determine total installation costs for that furnace can be tricky, especially if installation is not part of the original estimate from the contractor. When you tally capital costs for a heating system, make sure that you include all the costs of installation, including fuel tanks, special framing for heating system components, or perhaps even a woodshed if you are going to be burning wood.

Operating Costs

Operating costs include the amortization of your capital cost plus fuel expense and maintenance (including chimney cleaning, system tune-ups, and adjustments). The importance of the operating costs of a heating system far outweighs that of the capital costs, especially if you are planning to stay in your home for any length of time. One reason why electric baseboard heating was so popular in the 1960s was that the capital costs were very low. But the operating costs turned out to be a real shock—and they still are.

If you're considering a wood-burning heating system, your estimate of operating costs should include obtaining the wood, any related machinery and equipment for processing that wood, and (possibly) extra fire insurance. One of the problems with operating costs is that you don't generally find out about them until after you have moved in (or started using your new system) and begun to receive your fuel bills, usually one month at a time.

While one month's expense may not seem bad, if you add up the bills for one year, and then multiply that amount by ten or twenty years, the total can quickly chill your initial enthusiasm for your heating system. This might especially be true if you tried to save a few dollars by buying a cheap system: now you're stuck with a fuel hog, and possibly a maintenance headache as well, for as long as you own your house.

It is almost always cost-effective to spend more up front for a well-installed, high-quality, high-efficiency heating system that will save you money over the life of the system (it's even more cost-effective to design your home for maximum heat retention, thus lowering the size of the system that you will need in the first place).

Pop Quiz

When we look at the main elements of a heating system—heat source, distribution, controls, and heat storage—there's a key element that's missing from the picture, and that's *sustainability*. Okay, it's pop quiz time. Which of the following makes a home heating system sustainable?

a. Low operating costs
b. High-efficiency design
c. A programmable thermostat
d. Fuel
e. All of the above
f. I hate multiple-choice quizzes.

Now, I admit that low operating costs, high-efficiency design, and a programmable thermostat are all desirable things. But it's the *fuel* that is the main factor in determining whether a heating system is sustainable. You get a gold star if you picked "d."

Efficiency

Given the importance of choosing a high-efficiency heating system, you need to understand how heating-system efficiency is measured. First, how do we define efficiency? The **efficiency** of a heating system is the ratio between the amount of usable heat produced and the amount of potential energy in the fuel. Efficiency can vary from 0 percent or less (for some fireplaces) to as much as 90 percent for some high-efficiency furnaces. In heating systems, energy is lost in several places in the conversion and distribution process, and some simply goes up the chimney (if you use one).

The three main types of efficiency measures are **combustion efficiency**, which measures the efficiency of the combustion process; **steady-state efficiency**, which measures the system when it is running at its normal operating temperature; and seasonal or **annual fuel utilization efficiency**

(AFUE). Combustion efficiency is similar to highway mileage fuel consumption for your car, while AFUE is more like the average mileage your car gets across all driving conditions. That's because AFUE includes heating system start-up and cool-down and other losses associated with normal operation during the year. Steady-state efficiency is the ratio of the heat actually available for use in the distribution system to the amount of heat potentially available in the fuel. Since steady-state efficiency factors in heat lost up a chimney or smokestack, it's lower than combustion efficiency, but higher than AFUE.

The efficiencies of wood- and pellet-fired stoves, furnaces, and boilers can vary greatly, from 50 to 80 percent of available usable heat. Masonry heaters have higher combustion and thermal efficiencies than either woodstoves or fireplaces. A combustion efficiency of 90 percent is not uncommon for a masonry heater.

Efficiency measures for solar heating systems and heat pumps are expressed differently. Most solar collectors have a thermal performance rating based on characteristic all-day energy output measured in Btu. Heat pumps use two ratings to measure performance: one is the coefficient of performance (COP); the other is an energy-efficiency ratio (EER). With both of these ratings, the higher the rating, the better the heat pump. In addition, heat pumps have the highest operating efficiencies (up to 400 percent) of any of the heating appliances I describe.

Installation

When choosing a heating system, keep in mind the practical issues related to installing it. All heating systems can be considered for new construction, as long as the fuel they use is available in your area. However, some systems—such as passive solar and those that use ductwork to distribute hot air—may not be practical for all retrofit or renovation situations. Radiant-floor systems and some masonry stoves fall in the same category. While you probably *could* install almost any heating system in almost any home, some combinations just wouldn't make any financial sense.

Safety Issues

Do you have an infant or toddler crawling (or staggering) around your home who might bump into (or fall against) a hot woodstove? This is a crucial safety issue that often discourages many families from installing a woodstove as their primary (or secondary) source of heat. I know some people who have raised a family safely in a wood-heated home. However, I would be remiss if I didn't mention this concern.

Woodstoves also have the potential for creating a variety of fire hazards, which I will discuss in more detail in part 3. There are safety issues related to almost any heating appliance that uses an open flame. This relates to the phenomenon of **back drafting,** which occurs when excessive negative pressure in a house causes the combustion gases to spill back out of a heating appliance into the living space. This can be lethal, since the gases include carbon monoxide (CO), which is colorless, odorless, and highly toxic. Carbon monoxide is a potential killer and should not be underestimated. Back drafting can be the result of malfunctions in furnaces or badly designed chimneys, but the worst offenders are large-capacity exhaust fans (particularly downdraft cooktops in kitchens), especially in tightly constructed, energy-efficient homes. Installing a CO monitor, in addition to a standard smoke detector, is a good idea in any home that relies on combustion appliances for heating.

Maintenance

All heating systems (with the possible exception of passive solar) require maintenance. Most mechanical heating systems usually require quite a bit of maintenance in order to function efficiently and safely. The more complex the system, the more maintenance required. Filters need to be cleaned or changed; pumps or motors need to be checked; heat exchangers, burn pots, and flues need to be cleaned; and ash pans (in wood- and pellet-fired stoves, furnaces, and boilers) need to be emptied. You can take care of some simple tasks, such as emptying ash pans, yourself. Leave more complex tasks, such as servicing some components of a heat pump, to a specially trained technician.

Whatever system you eventually choose, be sure to keep the maintenance instructions for your heating appliance in a handy spot for future reference.

Warranties

Most heating appliances carry a warranty against defects in materials and workmanship. Warranties vary considerably among manufacturers and different types of systems. Read the fine print carefully, and be sure you understand the terms of the warranty before you plunk down your hard-earned cash. Also, check the reputation of the manufacturer of your system, possibly by talking to other homeowners who have installed the system you are considering. Comments from satisfied (or dissatisfied) customers are always more useful than a slick advertising brochure. It's a good idea to deal with a manufacturer that has been around for a while and that will probably be around for a while longer. This is important when you need replacement parts for your appliance. However, don't overlook a start-up company if it offers a truly revolutionary new product. Just be aware that you are taking more of a risk in terms of future availability of replacement parts, because the start-up may go belly-up.

Two Common Choices

Think of a heating system as a giant Lego set for adults. Like a Lego toy, a heating system consists of several parts, and there are many potential ways to put the parts together. A new add-on wood-fired boiler connected to an existing oil-fired hydronic heating system, combined with a new sunroom next to your kitchen, is an example of this mix-and-match approach. Keep this Lego image in mind throughout the book as I discuss heating system components in depth.

As I explained in chapter 1, the main parts of a heating system are source, storage, distribution, and controls. I will begin my discussion of heating system choices with the two main types of mechanical heat-distribution components found in most homes in the United States and Canada: forced hot air and radiant hydronic (hot water). In chapters 3 and 4, I will cover the wide variety of choices of heat sources and fuels for heating systems.

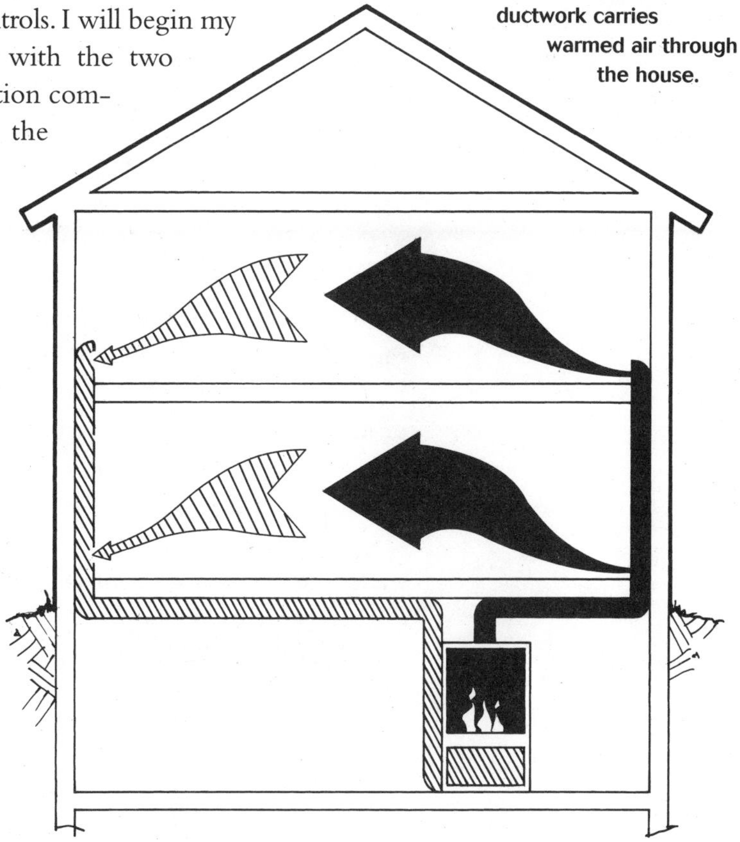

In a forced hot-air system, ductwork carries warmed air through the house.

Forced Hot Air

A **forced hot-air** (or warm-air) **system** has a heat source to heat the air and a ductwork system to distribute the heated air throughout the house. In addition to heat, the system can also provide air circulation, filtration, humidification, and cooling.

Heat Source

The heat source for a forced hot-air system is generally either a furnace or a heat pump. An electrically powered blower moves air through a **heat exchanger**, where it is heated. A heat exchanger is a device that transfers heat from one fluid (liquid or gas) to another while keeping the fluids separate. In this case, I'm talking about an air-to-air heat exchanger. (Different types of heat exchangers play a role in many systems discussed in this book, so it's a useful term to remember.)

Distribution

The blower distributes the heated air through a system of ducts (some systems may use natural convection, but this is not common). The first duct, called a **plenum,** is usually a large rectangular box made of sheet metal located on the hot-air (supply) side of the furnace. A cold-air plenum is located on the return side of the furnace. Smaller trunk ducts attached to the plenum feed even smaller branch ducts that connect to heating registers in individual rooms. Branch ducts and trunk ducts also run back from cold-air-return registers in rooms to the cold-air plenum on the furnace, completing the air loop. A well-designed hot-air system will have matched flow between supply and return air for all parts of the house.

In addition to sheet metal, some ducts are made of fiberglass duct board. Flexible branch ducts can be made of blanket insulation covered with a flexible vapor barrier and supported by a wire coil. The forced convection from the ducts generally blows hot air against exterior walls and windows in individual rooms.

Controls

Forced hot-air systems generally serve an entire house as a single heating zone. One thermostat located in a central part of the house (often the living room) controls the system. To control the temperature of individual rooms, homeowners must open or close individual hot-air registers. Newer, sophisticated hot-air systems can include two or more separate heating zones.

Other Factors

Hot-air systems are widely used, but some people don't like them. Since these systems blow a lot of air around, they tend to produce dust. Hot-air systems can also spread particles that cause problems for people with allergies. A clean-air filter can help reduce, but cannot eliminate, this problem.

The ductwork for a forced hot-air system takes up lots of space, especially in the basement, but also can be a design challenge in living spaces. If the system is also going to be used for cooling (with central air-conditioning and especially with a heat pump), the ductwork needs to be even larger. On the positive side, in solar-heated homes, hot-air systems can be used to move solar-heated air to parts of the house that do not receive direct sun. Some solar hot-air systems may even include a means of storing solar heat in a heat sink.

Because there are so many variables, an efficient and effective forced hot-air system should be designed and installed by professionals. This is particularly important because an improperly designed system can be noisy and inefficient. In a worst-case scenario, a badly designed system can even be dangerous if ducts leak or if negative air pressure is created in some parts of your home.

Hydronic (Hot-Water) Heating

The second main heating distribution strategy in conventional homes is radiant hydronic (hot water). I should point out that radiant heat can be produced by a wide variety of sources, including the sun, a woodstove, a masonry heater, or even a fireplace. However, most professionals in the heating trades are generally referring to steam or hot water produced by a boiler when they talk about radiant heating. I'll be using that limited definition in this chapter.

Why hot water? There are several good reasons. But first, let me clarify that hot water is not the heat source in a hydronic system. Hot water is an excellent medium for *transporting* heat from the source (usually a boiler) to the living space in your home. Water also acts as a storage medium, so hydronic heat tends to be more even than forced hot-air heat. And in solar-based (and some wood-fired) heating systems, water in a tank can be used

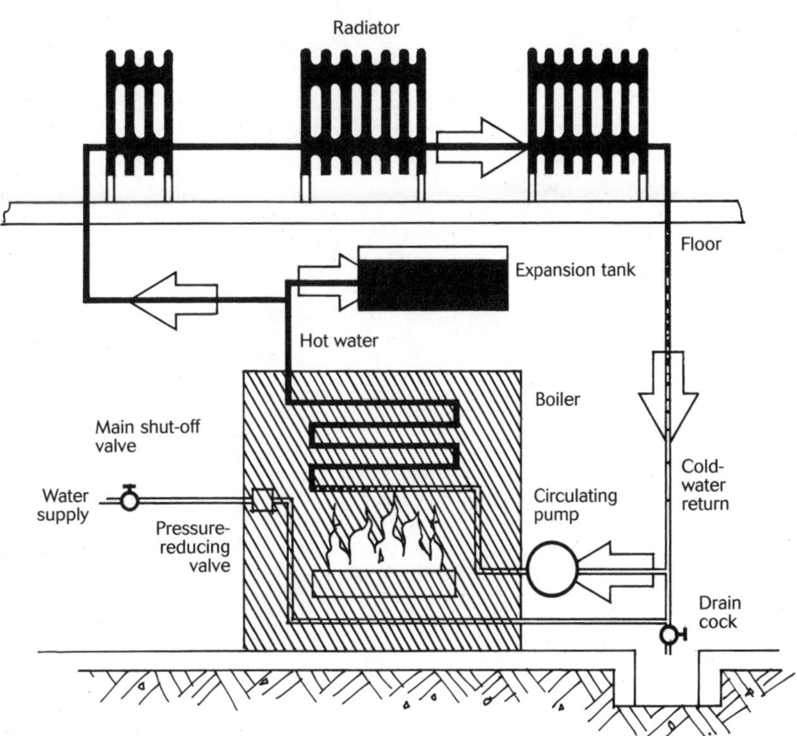

In a radiant hot-water heating system, water transports heat from the boiler to the living space through piping.

even more effectively for storing heat (refer to part 2 for more information on solar heating, and to chapter 12 for details about wood-fired central-heating boilers with storage tanks).

Heat Source

A hydronic system has a heat source (usually a boiler or heat pump but sometimes a solar collector) to heat the water and a system of pipes to distribute the heat throughout the house.

Distribution

A circulating pump distributes the heated water through the piping to a series of heat emitters in the living spaces and then back to the heat source for reheating, completing the loop. The **heat emitters** can be baseboard heating units, radiators, or radiant flooring.

Hydronic baseboard heaters are units that contain a pipe, usually with fins that transfer heat from the water to the air. Baseboard heaters make use of both convection and radiant heat transfer, creating a gentle flow of warm air, usually on exterior walls and especially under windows.

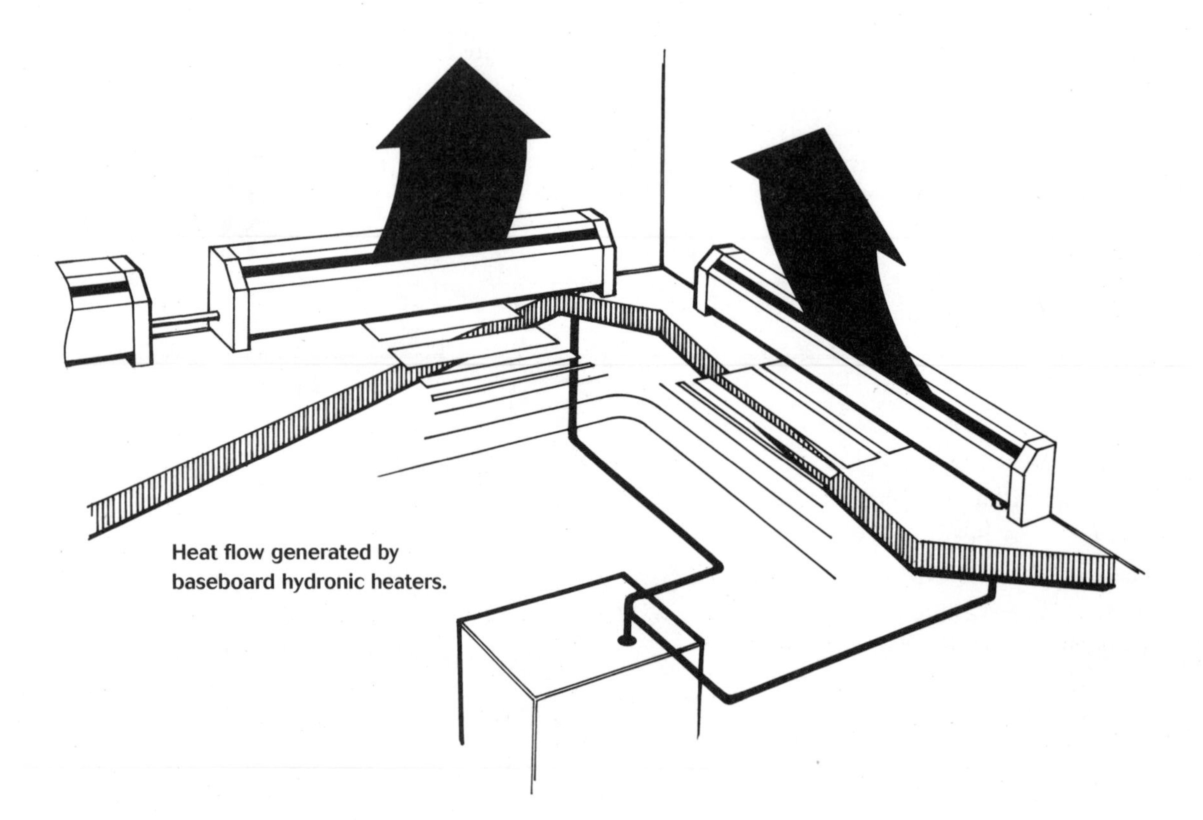

Heat flow generated by baseboard hydronic heaters.

The other common hydronic heat-emitting strategy is a radiant floor. Radiant floor heat dates back at least to Roman times. In its latest incarnation, however, a radiant floor system usually involves a pattern of pipes embedded in a concrete slab at grade level or in a thermal mass poured on top of a wood subfloor. The heat in the water in the pipes is transferred to the floor mass, which then becomes the heat emitter.

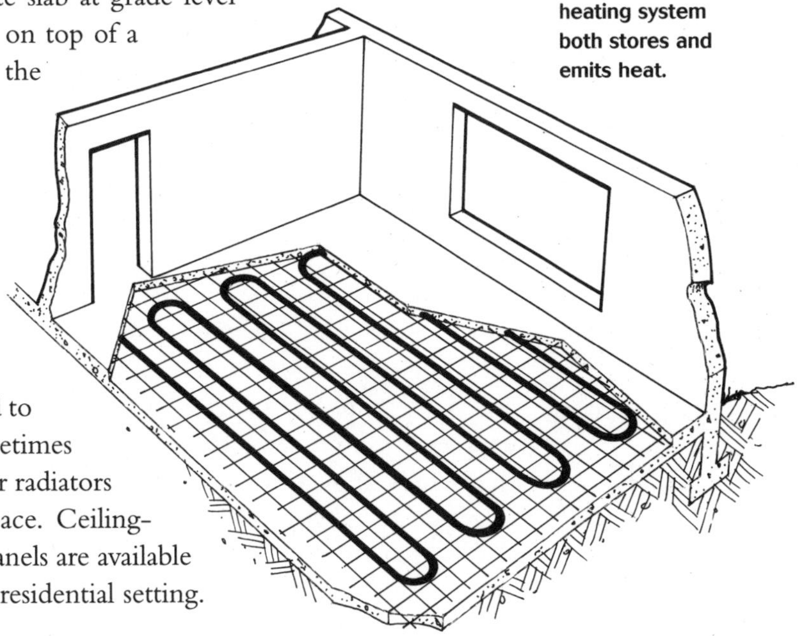

A radiant-floor heating system both stores and emits heat.

Radiant hydronic wall panels are yet another heat-emitter strategy that has been growing in popularity. These units are attached to wall surfaces and are sometimes used as replacements for older radiators that take up more floor space. Ceiling-mounted hydronic heating panels are available but are not widely used in a residential setting.

Controls

Controls for hot-water heating systems are generally more flexible and sophisticated than for hot-air systems. It's easy to establish separate heating zones for each level of your house or even for individual rooms.

Combination Systems

There is one variation on the hydronic heating strategy that deserves mention: the combination system (also known as an integrated system). This involves combining your home heating system with your system for heating domestic hot water. The heat source is generally your boiler, although solar panels can be used in some situations. Combination systems have the same advantages as a standard hydronic heating system, plus they yield plenty of hot water for baths, showers, washing dishes, and so on. Because the heating system is doing double duty, however, it needs to be sized accordingly. You'll find an energy savings overall, since the same heating appliance is supplying all of your hot water needs.

Other Factors

Hydronic heat is highly popular in cold northern climates. It offers steady, even heat and, in the case of radiant floor systems, comfortably warm floors (this is a particularly popular strategy for bathrooms). Radiant flooring in a living room can be a treat too, especially if you spend a lot of time sitting on the floor. Many newer hydronic systems use separate heat emitters for individual rooms; with these systems, you can tailor the heating strategy for each room. In addition, a hydronic heating system that is properly designed and installed can control both air and surface temperatures, delivering a more even mean radiant temperature in your home. Hydronic heating systems also have the advantage of being clean and quiet. Hot-water heat does not produce airborne particles the way a hot-air system does, and a hydronic system can be almost silent. Hydronic heating systems are relatively easy to install, and there's no need to find space for bulky air ducts. Hydronic systems don't supply humidity, but if you add a separate humidifier to your home, you can create a very comfortable environment. I've lived with a wide variety of heating systems over the years, and for overall comfort and convenience, I prefer hydronic heat. You will have to decide for yourself what type of system you prefer.

CHAPTER 3

Heat Sources

At the heart of every heating system is a heat source, and most heat sources involve some kind of heating appliance. The one exception is the sun. As I've already mentioned, the sun is the oldest and best source of heat there is. In fact, the sun is really the *only* source of heat—all of the renewable home-heating strategies we'll look at in this book are based either directly on solar energy, or indirectly on fuels derived from solar energy.

The Sun

Although primitive cultures sometimes used the sun for home heating (stone and adobe structures, for example, absorb and later emit solar heat), more "advanced" cultures like our own have mostly ignored the sun's potential until recently. We're finally realizing that solar heat is the simplest and least expensive heat source. There are two main strategies used to harness solar energy for home heating: passive and active.

Passive Solar

In a **passive solar** heating system, the sun is the heat source, while a house and its various components are the storage, control, and distribution systems. The siting and design of the house allow it to absorb solar heat in the winter and to avoid absorbing it in the summer. Passive systems operate for free and do not require electrical or mechanical devices to function. When included in the original design of a home, a passive solar heating system doesn't materially add to construction expenses, making passive solar the most cost-effective heating strategy for new construction. Passive solar heating systems can sometimes be retrofitted to existing homes.

Active Solar

In an **active solar** heating system, the sun is still the heat source, but a group of specially designed **solar collectors**—devices that harvest the sun's energy—is required. Water or air is heated in the collectors, and pumps or fans distribute the water or air through pipes or ducts to your living space. Active solar heating systems also have some kind of storage capacity to provide heat when the sun is not shining.

Active solar heating is a somewhat less desirable option than passive solar. Active systems can be expensive to install (from $10,000 to more than $18,000), require electricity to operate, and may not provide for all your heating needs (especially in cold, cloudy northern climates). A backup heater is usually necessary. On the plus side, it's almost always possible to retrofit an existing nonsolar-designed home with an active solar heating system.

Stoves

The stove is the oldest true heating appliance. Around 2,000 B.C., ancient Egyptians were using clay and brick ovens to bake bread. As I mentioned previously, stoves also were used to heat homes in Europe during Roman times. Then, between 1550 A.D. and 1850 A.D., Europe suffered a long period of extreme cold now known as the "Little Ice Age." As the effects of this climatic anomaly deepened, the supply of firewood in Europe dwindled, and there was a strong incentive to develop a more efficient heating appliance. The first real stoves for home heating were made of brick or tile. With the development of the simple cast-iron stove, home heating technology diverged and followed two main paths. Early brick and tile stoves evolved into sophisticated masonry heaters, while crude 15th-century cast-iron stoves evolved into the mind-boggling array of home heating stoves available today.

A six-plate cast-iron stove from the 1740s.

Courtesy of Antiquestoves.com

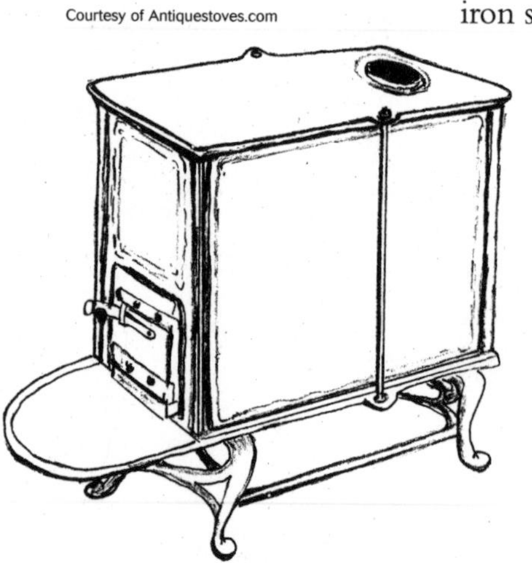

How They Work

Heating stoves aren't rocket science, and they still have the same basic design developed back in the 15th century. Stoves contain a combustion chamber where fuel is burned. The heat from that combustion is distributed into the living space primarily by radiation and sometimes by convection as well. Modern stoves usually include refinements such as baffles, primary and secondary combustion chambers, heat exchangers, dampers, and catalytic combustors. All of these refinements provide greater efficiency. Some stoves, especially pellet stoves, are equipped with a blower that distributes a portion of the heat by forced convection.

Other Considerations

Large heating stoves can heat an entire house, especially if the house is well insulated. Smaller heating stoves fall into the category of **unitary space heaters**, which means they are intended to heat only the room or area in which they are located. Installing a small stove can be a useful strategy if your home has a hard-to-heat room or addition. Some stoves can be equipped with a water jacket, which supplies domestic hot water. Depending on their design, heating stoves can burn a wide variety of renewable and nonrenewable fuels. If you are going to be away from your home during the winter for twelve to twenty-four hours at a stretch, a backup heater will be necessary; precisely how long a stove will be effective depends on its size and the fuel that it burns.

Most stoves (except pellet stoves) are manually controlled by dampers and air-inlet settings. Except for units equipped with fans and augers (pellet stoves), most stoves do not require electricity to function, which is a real plus in locations where power outages are frequent. Stoves can easily be used in new construction and most renovation projects. Heating stove prices vary from around $500 to $2,900, depending on size and features. Plan on $200 to $1,000 for installation costs, depending on your particular situation.

A modern wood-fired furnace heats air.

Courtesy of Energy King

Hot-Air Furnaces

The first successful central-heating system in modern times was introduced in 1835 in England. The system used hot air. In the late 1800s and early 1900s, there was a gradual shift away from wood and coal heating stoves in kitchens and parlors to central-heating systems in basements. The goal was to keep dust and dirt associated with wood and coal in the basement and out of the living space, while at the same time heating the entire house with just one appliance. The hot-air furnace is one type of heating appliance that evolved from that trend. Early hot-air furnaces were just a larger version of a wood or coal stove with a hot-air register added in the basement ceiling for heat distribution. Modern furnaces of all types are the most commonly used residential heating source in the United States.

How They Work

Like a stove, a hot-air furnace has a combustion chamber. The heat from the combustion chamber passes through a metal heat exchanger (sometimes several) where the heat is transferred to cooler air from the house. But unlike a stove, a hot-air furnace generally relies on a blower to move the heated air through a system of ducts to the rooms in your home (sometimes, the heated air moves by convection instead). Most hot-air furnaces are equipped with a filter that removes dust and allergens from the air. Often, the hot air in the furnace passes over a pan of water, thus providing humidification for your home as well.

Other Considerations

A basement or separate utility room is the best location for a hot-air furnace because of its size, reliance on a bulky ductwork distribution system, and noise level when operating. A hot-air furnace is the obvious heat source for a home that already has a forced hot-air ductwork distribution system. (If properly designed and installed, that ductwork can usually serve double duty for air-conditioning.) Depending on their design, forced hot-air furnaces can burn a wide range of renewable and nonrenewable fuels. With some fuels, a backup heater may be needed if you are going to be away from your home for more than a day during the winter (depending also on the size of your furnace).

Hot-air furnaces are controlled by a thermostat, which is usually mounted on a wall in the living space. Because they are normally equipped with fans and burners (and, in the case of pellet furnaces, mechanical augers), most hot-air furnaces require electricity to function, which can be a disadvantage in locations where power outages are frequent. Hot-air furnaces can generally be used in both new construction and renovations. A hot-air furnace of average size (100,000 Btu output) costs between $2,000 and $3,500 installed (cost of distribution system not included).

Boilers

The main difference between a furnace and a boiler is that a furnace heats air and a boiler heats water. The boiler is another outgrowth of the shift to central home heating. The first boilers for that purpose were developed in England in the 1850s. Early boilers heated water to create steam, which was then circulated through pipes to radiators in the living space of the home. Wood or coal was the fuel source for these boilers. Modern boilers are made of cast iron or steel. Cast iron is most common in residential hydronic systems, but steel boilers are also acceptable. These boilers are often much smaller than earlier models. Don't let the diminutive size fool you—these little units crank out plentiful heat with great efficiency.

A modern wood-fired boiler heats water.

Courtesy of Energy King

How They Work

Boilers heat water to temperatures between 140 and 180 degrees Fahrenheit. A pump circulates the heated water through a system of pipes to heat emitters located in the rooms of your home. After passing through the heat emitters, the water (now cooled) returns to the boiler, where it is heated again. This type of system generally requires an expansion tank to compensate for the varying pressure caused by the heating and cooling of the water.

Other Considerations

Although not as noisy as a hot-air furnace, a boiler is usually located in the basement or utility room. A boiler is the obvious heat source for a home that already has a hydronic distribution system, but a boiler can also be connected to a hot-air distribution system through a water-to-air heat exchanger. Depending on their design, boilers can burn a wide range of renewable and nonrenewable fuels. With some fuels, a backup heater may be needed if you are going to be away from your home for more than a day during the winter (depending also on the size of your boiler).

Boilers are controlled by one or more thermostats, usually mounted on the walls in the living spaces of your home. Because they are equipped with burners (and, in the case of pellet-fired boilers, mechanical augers), most boilers require electricity to function. Boilers can be used in new construction and most renovation scenarios. The cost of an average-size boiler runs between $3,000 and $5,000 installed (cost of distribution system not included).

Fireplaces

Picture the most inefficient, dirty, and problematic home heating device you can imagine. It couldn't be any worse than a fireplace. In particular, the traditional open English fireplace without doors is the most inefficient, and quite simply, it is not an effective home heating system. What's more, this type of fireplace tends to smoke up a house, especially during the early stages of the fire before a strong draft has been established in the chimney. Most fireplaces of this type probably have an overall efficiency rating in the minus column, since more hot air is generally sucked out of your home up the chimney than is added to your living space by the fire. Although sitting in front of an open fireplace on a cold, dark winter's night can be romantic, the romance cools pretty quickly at 30 degrees below zero.

An early English fireplace produced a lot of smoke—but not much heat.

Courtesy of Vance Smith

How They (Don't) Work

Fireplace operation has been set since about the 12th century. You kindle a wood fire in the fireplace: once the fire gets going (and stops smoking up your house), heat is distributed, mostly by radiation and partly by convection. Some fireplaces have fans and ducts that help distribute warm air by forced convection. The heat, such as it is, is controlled by the amount of wood you add to the fire. Fireplaces rarely add any heat to adjacent rooms unless forced convection heat ducts are used.

Other Considerations

In order to be even marginally effective, fireplaces and chimneys should be located inside the heated building envelope of your home, since a fireplace and chimney on an exterior wall will radiate heat out of your home. While the mass of a fireplace and chimney does offer some heat-storage capacity, most of this heat escapes up the chimney. Wood-fueled fireplaces may not be a viable option in urban areas due to lack of firewood supply and air pollution ordinances. At best, an unmodified open fireplace can only be considered a secondary, mostly ornamental heat source. A separate primary heating system will be required. On the plus side, except for fireplaces equipped with blowers, electricity is not required for operation. Fireplaces are generally more appropriate for new construction situations. A fireplace costs $2,500 or more. Happily, there are ways to improve a fireplace to make it a more viable and efficient heater (in chapter 13, I'll explain how).

Masonry Heaters

From the 15th century on, dwindling wood supplies in many parts of Europe sparked the search for more efficient stoves. Yet the real breakthroughs in this technology didn't come until the 17th and 18th centuries, when northern European monarchs ordered their stove craftsmen to devise more efficient heating stoves. Hoping to avoid royal displeasure, the craftsmen worked diligently, and the results were impressive. Masonry heaters with a variety of sophisticated designs became popular in Russia, Eastern Europe, Austria, Germany, and especially Scandinavia.

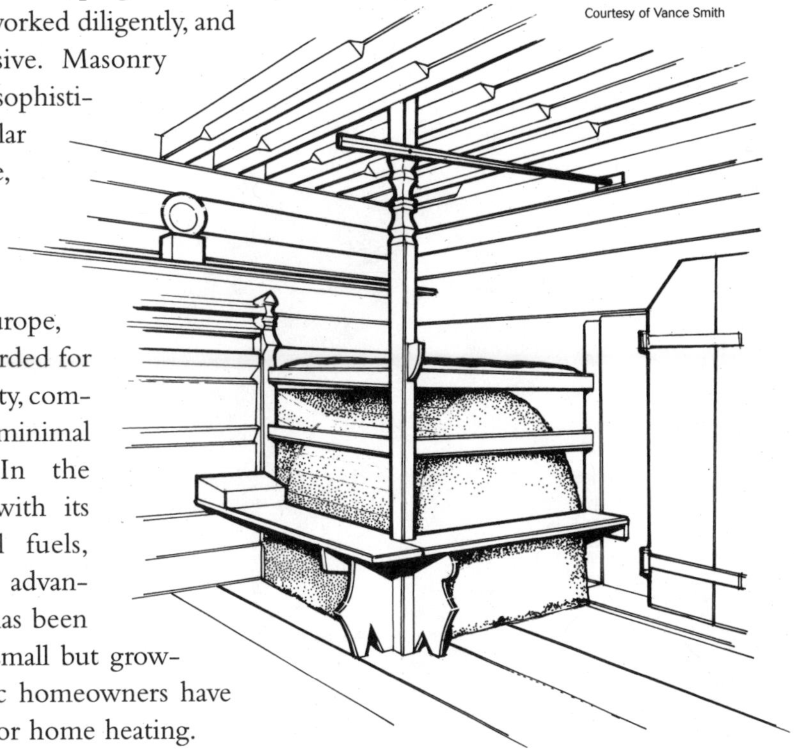

An early masonry heater like this was often located in villages in the Swiss or Austrian Alps.

Courtesy of Vance Smith

Masonry stoves are still popular in northern Europe, where they are highly regarded for their excellent heating ability, comfort, safety features, and minimal environmental impact. In the United States, however, with its heavy reliance on fossil fuels, recognition of the many advantages of masonry heaters has been slow. Recently, though, a small but growing number of enthusiastic homeowners have installed masonry heaters for home heating.

How They Work

Masonry heaters are designed for quick, intensely hot wood fires in their combustion chambers. The heat from these fires—which reach 1,300 degrees Fahrenheit or higher—passes through a series of baffled chambers and is absorbed by the large thermal mass of the heater. The heat then radiates slowly and gently into the surrounding living space. Masonry heaters are generally fueled once a day with bundles of small-diameter firewood (smaller in diameter than the wood used in a conventional woodstove) for quick, hot combustion. Heating efficiencies of up to 90 percent can be achieved with very low emissions. The amount of heat produced is controlled by the amount of wood burned.

Other Considerations

When properly designed and located, a masonry heater can be an excellent primary heat system, especially for a superinsulated home. A masonry heater can also act as additional thermal mass for a passive solar home. However, a backup heating device is generally required for times when you will be away from your home for more than a few days during the winter. Because they are extremely clean burning, masonry heaters are generally allowed in areas of the United States where other types of stoves and fireplaces are banned. Masonry heaters operate without electricity, an advantage in locations where power outages are frequent. Masonry heaters are best suited for new construction, but some designs can be added to existing homes. Masonry heaters cost around $10,000 installed; a variety of factors can drive the price up or down.

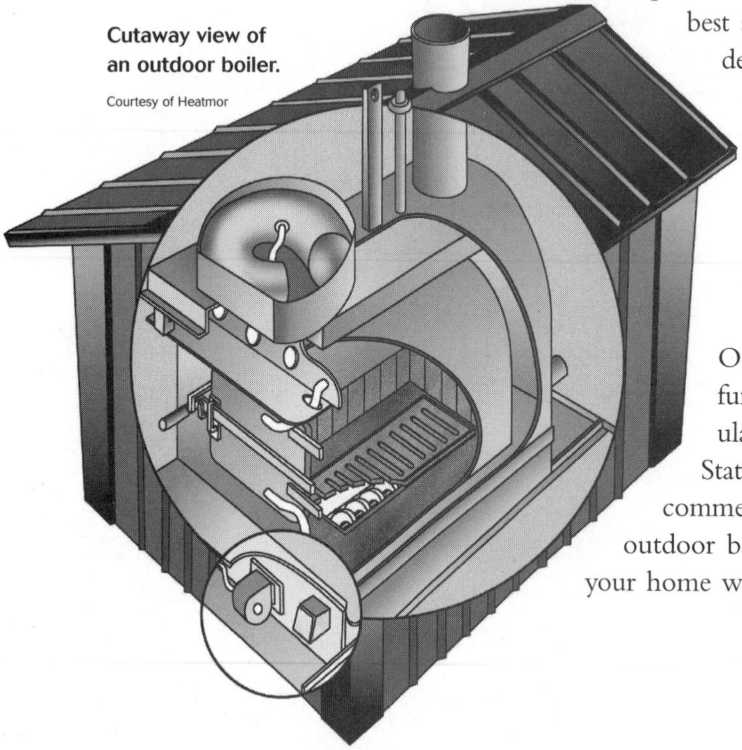

Cutaway view of an outdoor boiler.

Courtesy of Heatmor

Outdoor Boilers

Outdoor boilers (often erroneously called furnaces by their manufacturers) are popular in many rural parts of the United States and Canada, for both residential and commercial purposes. Developed in the 1980s, outdoor boilers offer an alternative way to heat your home with wood and other biomass. Outdoor boilers are weatherproof, and some brands even come equipped with an exterior covering that looks like a small storage shed, which protects the boiler's electrical and other components from the elements.

How They Work

In most designs, a water jacket surrounds the firebox of an outdoor boiler. Heat is transferred from the fire to the water in the water jacket; the heated water is pumped through insulated underground water lines to your house. The hot water then passes through a heat exchanger or directly into a variety of heat emitters; the water lines can be connected to your domestic hot-water tank as well. A properly sized outdoor boiler can heat not only your home, but also separate garages, sheds, workshops, barns, and swimming pools.

Other Considerations

Outdoor boilers are used as primary or secondary heat sources and can easily be interconnected with other types of boilers and furnaces. Because the combustion takes place outdoors, you won't have to contend with the dirt and dust normally associated with burning wood in the home. However, you will need to don boots, cap, gloves, and parka and trudge out into blowing and drifting snow in the middle of the winter to keep the fire going. Because they are located outside, these boilers are exempt from U.S. Environmental Protection Agency (EPA) emission standards. Due to their oversize fireboxes, many outdoor boilers are hard to regulate for a clean burn, and they tend to produce more air pollution than their indoor cousins. Outdoor boilers are not practical for most urban settings.

An outdoor boiler matches well with a home that already has a hydronic distribution system. Depending on their design, outdoor boilers can burn a wide range of fuels. A backup heater will be needed if you are going to be away from your home for much more than twenty-four hours during the winter. Outdoor boilers are controlled by a thermostat. Because they are usually equipped with pumps and other electrical controls, most outdoor boilers require electricity to function. Since they are located outside of the home, outdoor boilers are appropriate for both new construction and renovations. Outdoor boilers cost between $4,500 and $5,500. Installation costs range from $1,200 to $1,500 (see chapter 12 for more on outdoor boilers, including some additional disadvantages).

Heat Pumps

Heat pumps have been around since the early 1900s, and you probably already have one in your home, disguised as a refrigerator. A heat pump is a mechanical device, with a compressor as its main component, which heats or cools a building using one of several heat sources. Heat pumps for residential heating purposes were first developed in the 1970s. There are two main types of heat pumps: air source and ground source. Initial installation costs for heat pumps (especially ground-source heat pumps) are generally high, but operating costs tend to be low. Heat pumps have operating efficiencies well over 100 percent—sometimes as much as 400 percent! Heat pumps can be installed in almost any climate; some types work better in milder climates, while other types work better in colder climates.

How They Work

In their heating mode, heat pumps force heat to move from a cool space to a warm space (opposite to the natural movement of hot to cold). Using an electrically powered compressor, heat pumps can transfer heat from a wide range of heat sources in the environment to the living space of your home. Those sources include air, water, the earth, and even the sun. Heat pumps can also provide domestic hot water, dehumidification, air filtration, and cooling during the summer. When the heat-pump process is reversed, it moves warm air out of your house (instead of in). This ability of a heat pump to double as a cooling device is a real advantage, especially in warmer climates.

Ground-source heat pump components.

Courtesy of WaterFurnace

Other Considerations

Since there is no combustion involved, heat pumps tend to be very safe, with virtually no risk of fire or combustion gases escaping into your home's living space. Heat pumps are generally connected to hot-air distribution systems but can be used with some hydronic heating systems as well.

Heat pumps are controlled by a thermostat. Because they are equipped with a wide range of electrical devices and controls, heat pumps require electricity to function. Heat pumps are particularly suited for use in new construction, yet can also be installed in some retrofit situations. Heat pumps cost between $3,500 and $15,000 installed (air-source heat pumps are less expensive than ground-source heat pumps). Prices vary depending on a wide range of factors.

Fuel Cells

The fuel cell is not primarily a heating appliance; it produces electricity. Heat is the by-product of the production of electricity. That "waste" heat could be used to warm domestic hot water. Under ideal circumstances, the waste heat could be a supplementary heat source for a highly insulated passive solar home in a relatively mild climate.

There has been a lot of buzz about the potential for fuel-cell technology to revolutionize the energy industry. The key word here is *potential*. The main problem at present is that most fuel cells run on fossil fuels. The hope, eventually, is to power fuel cells with hydrogen that has been produced by renewable energy sources, but that vision has not yet materialized. While some fuel cells are presently capable of running on hydrogen, the lack of a large-scale system for hydrogen generation and distribution makes the widespread adoption of fuel cell technology problematic.

I'm enthusiastic about fuel cells for use in vehicles, but clean, renewable fuel sources for hydrogen production need to be developed before the fuel cell reaches its full potential. When that happens, incorporating a fuel cell into a renewable home heating system might make sense.

CHAPTER 4

Fuels

Without fuel, your heating system is just a collection of cold metal parts (or, in the case of a fireplace or masonry heater, cold bricks and mortar). Add fuel to the picture, and then you have a functional heating system. However, even if you have the best designed, most efficient, top-of-the-line heating system available, it doesn't qualify as a sustainable system if it burns a nonsustainable fossil fuel such as oil, gas, or coal. Unfortunately, the use of fossil fuels and electricity (which is generated mostly by burning fossil fuels) accounts for about 95 percent of home heating in the United States. It's clear that we have an enormous problem with unsustainable home heating practices—especially with fossil fuels.

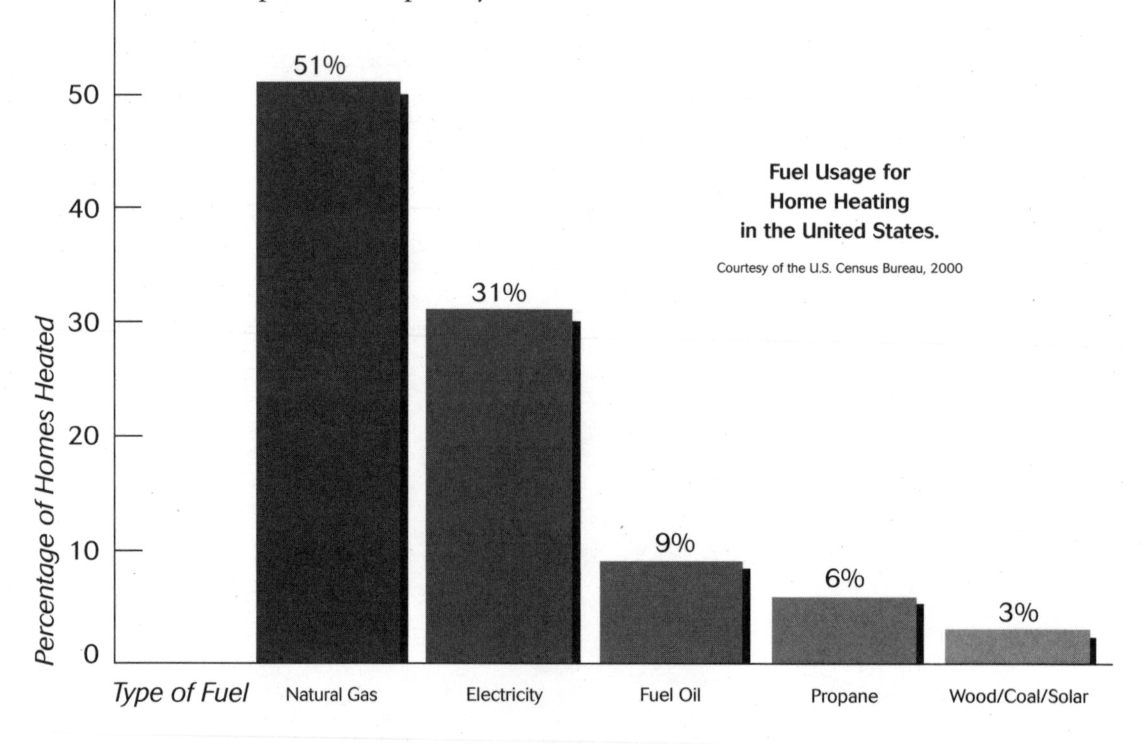

Fuel Usage for Home Heating in the United States.

Courtesy of the U.S. Census Bureau, 2000

The Problems with Fossil Fuels

Fossil fuels are energy-rich substances formed from the long-buried remains of prehistoric life. These fuels mainly consist of hydrocarbons, which are compounds composed of hydrogen and carbon. Fossil fuels are nonrenewable because it takes almost infinitely longer to produce them than it does to consume them.

Burning fossil fuels causes significant health and environmental hazards. In addition to causing air pollution, burning coal and oil also produces carbon monoxide and tiny particles that have been linked to lung cancer and other health hazards. Other by-products cause urban smog and acid rain. Fossil fuel combustion also produces enormous amounts of carbon dioxide, one of the primary greenhouse gases that lead to global warming. Although natural gas burns more cleanly than other fossil fuels, it still gives off greenhouse gases when burned.

Because of their reliance on fossil fuels, heating systems in the United States emit one billion tons of carbon dioxide and substantial amounts of other pollutants annually. These gases, along with emissions from vehicles and industry, contribute to global warming. (Most scientists agree that global warming is actually taking place and that our burning of fossil fuels is a major contributor.)

The burning of fossil fuels has raised the concentrations of carbon dioxide in the atmosphere by about 30 percent since the industrial era began. The levels of carbon dioxide are now at their highest point in 160,000 years, according to recent studies. If this trend continues, higher average global

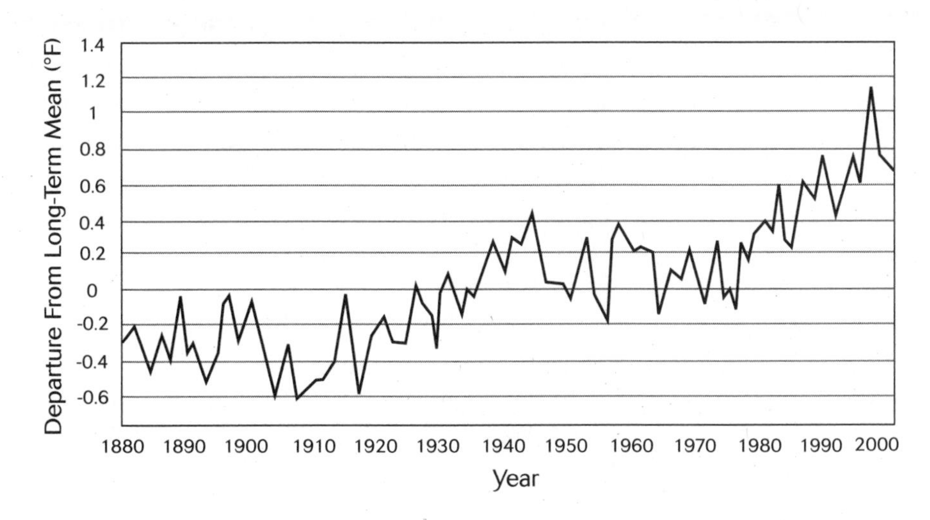

Global temperature increase has been dramatic since 1980.

Courtesy of U.S. National Climatic Data Center, 2001

temperatures are almost certainly going to follow: between 3 and 11 degrees Fahrenheit by 2100, according to the Intergovernmental Panel on Climate Change. The average global temperature in 2002 was the second highest on record since global record-keeping began 140 years ago (the highest was in 1998), according to the World Meteorological Organization.

Temperatures are rising and the Earth is growing hotter faster than at any time in the past. Nine of the ten warmest years in the last four decades have occurred since 1990, and temperatures are now rising three times faster than in the early 1900s. Rising temperatures can lead to more severe and unpredictable weather patterns, droughts, wild fires, floods, crop failures, rising sea levels, and a host of associated disruptions and disasters to humans and other life forms.

The mining and extraction of fossil fuels, along with their transport and processing, also causes environmental damage from oil spills, refining operations, water pollution, and ground pollution. There are serious political and military implications of our overreliance on oil imported from the Middle East, which were graphically brought home with the destruction of the World Trade Center in New York City on September 11, 2001, and the subsequent military campaigns in Afghanistan and Iraq. For all of these reasons, we need to eventually eliminate our use of fossil fuels.

Renewable Energy for Home Heating

Shifting from our current reliance on fossil fuels to the use of renewable sources of energy is a tall order. But every little bit helps, and changing the way we heat our homes is an important part of the larger initiative. Heating your home with renewables is a particularly attractive strategy because it's something that you can do yourself, right now; you don't have to wait for the government to act. Switching to renewable fuels may involve some modifications to your heating system, your home, and possibly your daily routine. There are many options, and each one has relative advantages and disadvantages.

The Sun

Technically, the sun isn't a fuel; it's an energy source. But for the purposes of this chapter, let's consider it as a fuel. From a human perspective, the sun is an inexhaustible source of energy. Enough solar energy strikes the Earth in one hour to power all human activities for an entire year. Much of this energy turns into heat, but plants absorb some of it as light energy. The plants, in turn, transform this solar energy into chemical energy through photosynthesis. All of the renewable fuels we'll be looking at in this chapter are ultimately derived from photosynthesis, except for geothermal.

Under the right conditions, the sun can be used as a "fuel" for both space heating and domestic-hot-water heating. However, use of the sun requires careful home design and becomes more problematic the farther north you live. Especially in most northern locations, some form of backup heat is required.

Solar Pros and Cons

Pros	Cons
Infinite supply	Needs to be part of basic house design for maximum effect
Nonpolluting	Not always available
Brightens and warms a home on cold, dark winter days	Less practical in northern latitudes
Free	Requires thermal storage to be effective

Wood

Wood is the oldest true heating fuel. Our earliest ancestors used wood for cooking and home heating for hundreds of thousands of years. But in the late 19th century, wood fell out of favor and was replaced by fossil fuels. Since the oil embargo in 1973 of the Organization of Petroleum Exporting Countries (OPEC), however, wood has once again become a viable home heating fuel in the United States (in many other parts of the world, wood never stopped being a fuel).

Heating with wood does not contribute to global warming as long as it is part of a managed resource cycle that includes replanting trees. The carbon dioxide that is released when wood is burned is taken up by the replacement trees that grow in a sustainably managed forest. This cycle can be repeated indefinitely without increasing atmospheric carbon. The same rule applies to all "biomass" (wood, wood pellets, vegetation, grains, or agricultural waste) used as a fuel or energy source.

If you are considering heating with wood, find a reliable long-term source of firewood in your area. This may be difficult, especially in many urban locations, and could cause you to choose a different fuel source instead.

Check local regulations too: concerns about air pollution caused by smoky wood fires have prompted some municipalities and states to restrict the use of wood-burning appliances. You will also need a place to store your firewood.

The heat production of wood varies, depending on moisture content and species. Most hardwoods typically produce around 29 million Btu per cord, while softwood only puts out around 17 million Btu per cord.

Wood Pros and Cons

Pros	Cons
Fairly low operating costs	More dangerous than some fuels if heating appliances are not properly installed or maintained
Can be used in a wide variety of heating appliances	Storage space and heavy physical labor required
Generally does not require electricity	Creates indoor dirt and dust problems
Low environmental impact if sustainably harvested wood is used and if it is burned responsibly	Requires a properly maintained chimney
	Can cause air-quality problems both indoors and outdoors
	Combustion gases can spill into the living space during a malfunction
	May increase fire insurance premiums

Pellets

Wood pellets are a relatively new heating fuel developed following the OPEC oil embargo. Pellets for home heating are generally made from sawdust and ground wood chips, which are waste materials from processing of trees for furniture, lumber, and other products. The productive use of material that would otherwise be wasted is a strong argument for using pellets for heating purposes, but there are other good reasons, too. Because of their design, most pellet-fueled heating appliances burn fuel steadily and very efficiently, with extremely low combustion emissions. Pellet-burning appliances need to be vented but don't require a standard chimney.

Heating costs for pellets are generally lower than for electricity and propane and about the same as for fuel oil and natural gas. Bags of pellet fuel are easy to handle and store; far less daily

Wood pellets, generally made from sawdust and ground wood chips, offer numerous advantages as a home heating fuel.

Courtesy of Amber Rood

Pellet Pros and Cons

Pros	Cons
Competitive operating costs	Storage space and some physical labor required
Easy to store and handle	Most pellet appliances require electricity
Burns cleanly	Combustion gases can spill into the living space during a malfunction
Low environmental impact	
Steady heat	
No chimney is required	

labor is required than for tending a woodstove. A pound of wood pellets produces about 9,000 Btu; this is equivalent to 18 million Btu per ton.

Corn

Burning corn in stoves and furnaces for home heating was a popular strategy during the Great Depression, especially in the Midwest, where millions of nearly worthless bushels of corn were readily available. It didn't take desperate farmers long to figure out that their corn was worth more as a heating fuel than as a feed commodity. They also discovered that burning corn produced steady, even heat and was far less labor intensive than burning wood. Since then, every time the costs of heating fuel and natural gas have exceeded the price of corn, farmers renew their enthusiasm for heating their homes (and outbuildings) with corn. And the same general strategy works for wheat, barley, rye, sorghum and even soybeans; when their prices fall low enough, they can be economically burned as fuel.

A pound of corn produces about 7,000 Btu (14 million Btu per ton). That's less heat output than wood pellets produce, but corn can be much less expensive than pellets. When the market price for grains is low and the

Corn Pros and Cons

Pros	Cons
Cheapest renewable fuel (when grain prices are low)	Requires a special heating appliance
Steady, even heat	Most grain-burning heating appliances require electricity
Burns cleanly	Storage facilities and some physical labor required
Low environmental impact	Stored corn (and other grains) attracts rodents and insects
Generally does not require a standard chimney	

price for heating fuels is high, corn can be the cheapest home heating fuel available—about half the price of wood pellets and two-thirds less than propane.

Biodiesel

Wouldn't it be nice if someone could develop a renewable, biodegradable fuel that could power vehicles, heat homes, cut dangerous emissions, assist farmers, and help to end our dependence on imported oil—all at the same time. Guess what? Someone already did, over a century ago. Rudolph Diesel, a German engineer, used peanut oil to power one of his engines at the World Exposition in Paris in 1900. Diesel's engine, which now bears his name, was originally designed to run on a variety of fuels, especially vegetable and seed oils. Diesel's use of peanut oil as a primary fuel source never became popular because of the subsequent market success of petroleum-based fuels. However, Diesel's idea is still as valid today as it was over one hundred years ago.

In the 1970s, in response to the OPEC oil embargo, petroleum makers extended limited fuel supplies by mixing gasoline with ethanol from fermented corn; it was called gasohol. More recently, Diesel's original idea of using vegetable oils as a fuel source has been revived with the development of biodiesel. **Biodiesel** is a renewable diesel fuel substitute that is easily made, through a simple chemical process, from virtually any vegetable oil, including (but not limited to) soybean, corn, canola, cottonseed, peanut, sunflower, and mustardseed. Biodiesel can also be made from recycled cooking oil or animal fats. There are even some promising experiments on the use of algae for making biodiesel. The possibilities seem endless.

Biodiesel readily biodegrades, so it is far less polluting than its petroleum-based alternative. It's even more biodegradable than sugar and ten times less toxic than table salt. Best of all, biodiesel is renewable. While fossil fuels took millions of years to produce, biodiesel sources can be created in just a few months. In a properly managed system, there is no net emission of carbon dioxide as a result of burning biodiesel, because carbon dioxide uptake by plants being grown for future biodiesel supplies offset the carbon dioxide released. It seems too good to be true, but, remarkably, it is true.

Until recently, biodiesel has mainly been promoted as a fuel for diesel-powered vehicles, but biodiesel can also be used as a fuel additive in a standard oil-fired furnace or boiler. I first heard about using biodiesel as a fuel additive for home heating when I attended SolarFest 2001 in Middletown

Springs, Vermont. I decided to try the technique myself. In November 2001, shortly after the fuel tank in our basement was filled with Number 2 fuel oil, I cautiously added about five gallons of biodiesel to the tank. (I waited until the tank was filled so I would know approximately what proportional mix I was creating, but I also held off a few days after the fuel delivery so I wouldn't accidentally overfill the tank.)

The biodiesel came from Dog River Alternative Fuels of Berlin, Vermont, and was made from recycled cooking oil supplied by the New England Culinary Institute. (I didn't climb up on the roof to check whether the combustion gases coming out of our chimney smelled like french fries. Diesel vehicles running on recycled-cooking-oil-based biodiesel often exhibit this quirky phenomenon.) I started the experiment with a modest amount because, among its many properties, biodiesel is also a solvent. It tends to dissolve the sludge that often coats the insides of old fuel tanks and fuel lines, and I thought that this might cause a clogged fuel filter or burner head. But our old 1956 American Standard oil-fired boiler in the basement ran without incident after I added the biodiesel. As the heating season progressed, I gradually upped the percentage of biodiesel after every fuel oil delivery until we were burning a B10 blend (10 percent biodiesel, 90 percent fuel oil), with no adverse effect whatsoever on the heating system. I could hardly believe

Biodiesel Pros and Cons

Pros	Cons
Will burn in virtually any oil-fired furnace or boiler	Tends to be more expensive than regular heating oil
Reduces harmful stack emissions	May not be readily available in all locations
Simple conversion process	May cause clogged fuel filters or burner heads in older systems
	B100 (100 percent) biodiesel will soften rubber hoses and gaskets and may corrode copper or brass fittings
	Combustion gases can spill into the living space during a malfunction
	Needs to be stored in an indoor tank in cold climates
	Oil-fired heating appliances generally rely on electricity to function

how simple the procedure was: no new heating appliance was required, and no retrofitting. All I had to do was add the biodiesel to my indoor fuel tank (biodiesel will gel if stored outside in extremely cold weather).

I understand that some people have successfully used B100 (100 percent biodiesel) to heat their homes, so biodiesel's potential for reducing petroleum-based fuel usage is considerable. For example, if everyone in the Northeast used just a B5 blend, it could save 50 million gallons of regular heating oil per year, according to officials at the USDA Beltsville Agricultural Research Center in Beltsville, Maryland. The center has been heating its buildings successfully with a biodiesel blend since 1999. "Using biodiesel offers an opportunity to reduce emissions, especially particulate matter and hydrocarbons, and that's a great advantage," says John Van de Vaarst, deputy area director, who is responsible for facilities management and operations at the station. "I think it's an obvious strategy to help clean up the environment and reduce our dependency on foreign oil."

I agree. There is enough biodiesel feedstock in the United States to produce 1.9 billion gallons of biodiesel per year. Another 5 to 10 billion gallons could be made from mustard seed, and billions more could be made from algae. Biodiesel is a simple, proven technology that, along with other renewable fuels and conservation strategies, could end dependence on foreign oil and dramatically improve air quality nationwide.

Geothermal Energy

Normally, geothermal energy refers to the heat of the Earth's interior. Visions of gushing geysers and hot mineral springs bubbling with sulfur come to mind. In some locations, geothermal energy has been tapped to generate electricity and heat buildings (in some countries, even to heat entire cities). But the term "geothermal" is also used more broadly in relation to heat pumps. **Geothermal** is a descriptive term for stored

Geothermal Pros and Cons

Pros	Cons
Infinite supply	Electricity is needed
Nonpolluting (except for the generation of electricity)	Requires an expensive special heating appliance
Safe; noncombustible; no risk of escaping fumes	May need an abundant source of water for some applications
Does not require a chimney	
Can be used for both heating and cooling	

energy, the result of solar energy being absorbed by the ground, ground water, or a body of water (rather than by plants). The definition of geothermal is sometimes stretched even further to refer to solar heating of the air. It's these broad definitions of geothermal that I use in this book. Since the sun is the source of geothermal heat for heat pumps, geothermal is essentially a free fuel that is completely renewable. But you need to have a heat pump in order to tap geothermal heat—and electricity to run the heat pump.

Electricity

Like the sun, electricity isn't a fuel; it's a form of energy. But no discussion of heating fuels would be complete without mentioning electricity. In the 1960s, electricity—especially nuclear-generated electricity—was touted as the clean, cheap, modern energy of the future. Residential electric heating was viewed as a key part of that rosy picture. Sadly, nuclear power didn't live up to the overoptimistic publicity. Neither did electric heat.

Heating with electricity turned out to be at least two times more

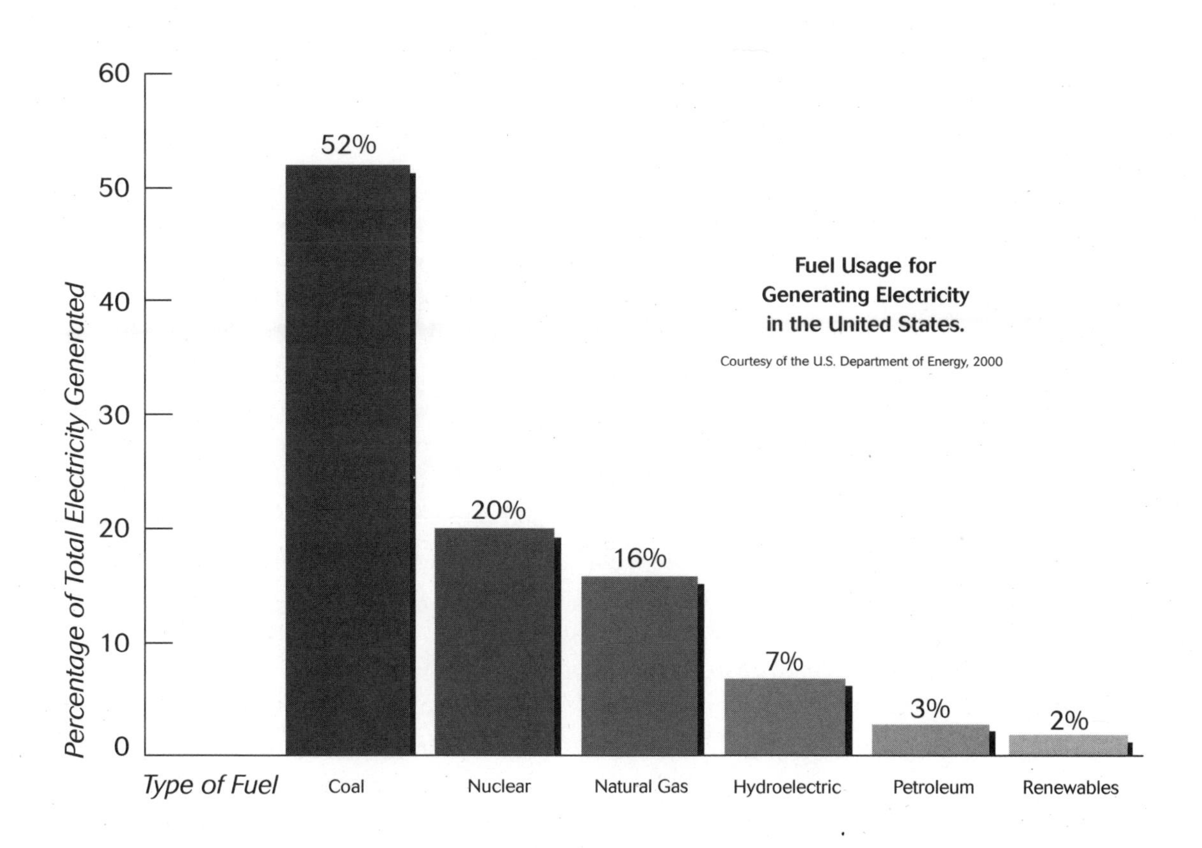

Fuel Usage for Generating Electricity in the United States.

Courtesy of the U.S. Department of Energy, 2000

expensive than most other methods. In addition to being outrageously expensive, electric heat is bad for the environment, because electricity is generated primarily from nonrenewable fossil fuels and also from nuclear power plants. I include nuclear power in the same unsustainable category as fossil fuels because no one has ever figured out a safe way to store the incredibly dangerous waste materials that result from nuclear power generation. The nuclear power industry loves to advertise the fact that their facilities don't produce greenhouse gases, but they fail to mention the nuclear-waste dilemma.

It is possible that, as the generation of electricity gradually shifts to renewable sources, electricity will eventually be considered a more renewable form of energy. That change is still a long way off, and until it happens, I can't recommend electric heat as a primary renewable home heating strategy. Electricity does, however, play an important role in the operation of the fans, pumps, burners, thermostats, and other electrical devices contained in many heating systems. Also, you may want to consider electricity as the power source for a secondary or backup system—as long as you won't need to use it very often.

Fuel Cost Comparison

Now that we've looked at the relative advantages and disadvantages of renewable fuels, let's consider their relative costs. Making simple cost comparisons between fuels is, unfortunately, not simple because units of measure for fuels aren't consistent. To further complicate the picture, fuel prices vary widely from one region to another and from one season (or even month) to another. As fuel prices (especially fossil fuel prices) become more and more volatile, comparisons are going to become even more difficult. Nevertheless, there are overall patterns that seem to hold true much of the time.

Fuels tend to fall into three main price ranges that I'll call "high," "medium," and "low." Electricity and propane gas are at the high end, with electricity by far the most expensive in most regions. In the middle range are pellets, natural gas, and heating oil (B20 biodiesel is a little more expensive than regular heating oil, but still falls in the middle range). At the low end are coal, softwood, hardwood, and corn.

If you eliminate nonrenewable fuels (electricity, propane, heating oil, natural gas, and coal), that leaves pellets, biodiesel, wood, and corn in descend-

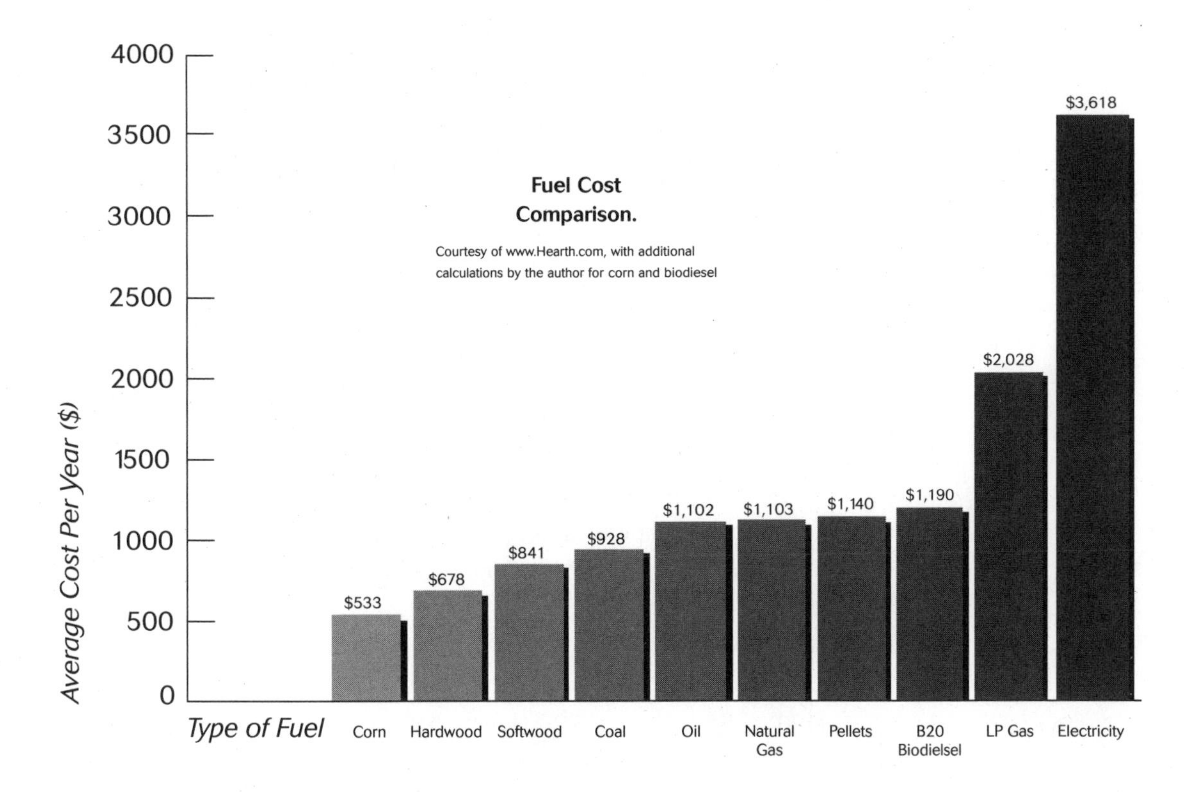

ing order of cost. Solar and geothermal are in categories of their own, but ground-source geothermal heating and cooling is typically 30 to 60 percent less expensive to operate than other conventional systems.

The best approach to follow for fuel comparisons, however, is to make your own. That way you will be sure your figures are current and appropriate for your location. The Office of Energy Efficiency and Renewable Energy of the U.S. Department of Energy has a useful formula for this purpose. You can use this formula to estimate the costs of producing 1 million Btu of heat using different heating appliances and fuels. To do this, you need to know the efficiency of the appliance and the unit price of the fuel (check with your utility or fuel supplier for the unit price of the fuel in question). Here's the formula:

> Energy cost = cost per unit of fuel ÷ (fuel energy content × heating system efficiency)
>
> Energy cost is expressed as dollars per million Btu.
>
> Cost per unit of fuel is expressed as dollars per gallon, pound, etc.

Fuel energy content is expressed in millions Btu per fuel unit.
Heating system efficiency is expressed as a decimal.

To use the formula, be sure to use a decimal equivalent for the appliance heating efficiency. Also, to determine the fuel energy content, you must *first* convert the Btu content of the fuel per unit to millions of Btu per unit by dividing the fuel's Btu content per unit by 1,000,000. For example, to determine the fuel energy content of electricity, divide 3,413 Btu per kilowatt-hour (kWh) by 1,000,000, which equals 0.003413 million Btu/kWh.

Here's how the energy cost of heating oil might be figured using this formula:

Oil (in central heating system) cost = \$1.25 per gallon ÷ [0.14 million Btu per gallon × 0.80 (efficiency)]
The answer is \$11.16 per million Btu.

For the mathematically challenged, happily, there is an even easier way to compare fuel costs for most fuels—an on-line fuel cost calculator. If it hasn't disappeared into cyberspace, you'll find the calculator at www.hearth.com/fuelcalc/findoil.html. The calculator not only computes the cost of producing one million Btu but also generates a comparative chart for average home heating costs per year. Make sure you enter your current local fuel prices as well as the efficiency figures for the heating appliance(s) you are comparing. The calculator section for oil can also be used for biodiesel, since the heat output is virtually identical; just enter the per gallon price for biodiesel (or a biodiesel blend) instead of oil.

There is also an on-line energy-selector Web site for corn hosted by Penn State University (located at http://energy.cas.psu.edu/energyselector/) that shows the fuel-switching point between corn and most other fuels (but you can also use it to compare the switching point for any of the fuels listed in the calculator). It's important to remember, however, that fuel price should not be the sole measure for selecting a heating appliance, since there are many other factors involved—not the least of which is sustainability.

PART TWO

HEATING WITH THE SUN

CHAPTER 5

Passive Solar

In 1974, I began construction on my first house. It wasn't really a solar home. The long dimension of the house ran north to south, and the roof surfaces were oriented east and west—a classic example of bad solar design. However, I did include some modest passive solar elements. I installed more windows in the south-facing gable end to take advantage of natural light and solar heat during the day. I put in only one small window at the north end, to minimize heat loss during the winter. But I made no effort to absorb, store, and use the valuable solar energy that streamed through the south-facing windows during the winter. The design was not bad for a first attempt, but it could have been much better.

Eight years later, after a move to a remote location in Lincoln, Vermont, I had another opportunity to design a house for myself. This time I made a serious attempt to incorporate more passive solar elements into the design. The roofline still ran the wrong way, but I included lots of large windows and several sliding glass doors on the first floor in an attempt to take better advantage of solar gain on the south-facing gable end. Because of thin soil cover, I opted to erect the structure on a 4-foot-deep "frost wall" foundation, rather than a full basement. Truckloads of crushed stone were spread on top of rigid foam insulation in the south end of the foundation area and then covered with a poured concrete slab. The concrete slab on the south end of the house was then overlaid with tiles of dark Vermont slate, which absorbed heat when the sun shone on them during the winter. That heat was then transferred to the concrete slab and the crushed stone beneath, which acted as a heat sink. At night, the heat would radiate gently back into the living space from the floor. We burned about four cords of wood every winter in a Vermont Castings stove (our primary heat source), and the house was light, welcoming, and pretty cozy most of the time. Although it wasn't a total passive solar design, it certainly was an improvement over my earlier house.

The Potential of Solar

Even in my first attempt at house building, I was aware of the importance of the sun (although I failed to take full advantage of its potential). The sun is a huge source of energy that dwarfs our limited supply of fossil fuels. The world's entire petroleum resource, for example, is estimated to be approximately one million terawatts (one terawatt equals one trillion watts), which roughly equals the amount of solar energy that reaches the Earth *every day.* While it took billions of years to produce fossil fuels, we're already beginning to run out of them after only a few hundred years of use. It's clear that making use of the direct energy from the sun is a more sensible and sustainable strategy in the long run, especially since it's free.

About one million Americans currently live in solar homes, and most of them are enthusiastic supporters of solar energy. But not all of them heat their homes with the sun. That's because there are many ways to harvest and use solar energy. Photovoltaic (PV) modules, for example, can turn solar energy into electricity. Using PV modules for your electrical needs works well, especially if your needs are relatively modest. However, heating your house with electricity is generally not a cost-effective strategy; trying to install enough PV modules to produce sufficient electricity to heat your home would be prohibitively expensive. However, a PV-powered circulating pump on a closed-loop active solar hot-water heating system, or a solar-assisted heat pump are two systems where photovoltaics could be used in a more cost-effective manner (see chapters 7 and 21 for details). On the other hand, heating your home directly with the sun's rays is much simpler and even more cost-effective.

In this chapter, we'll discuss passive solar heating systems; chapters 6 and 7 cover active solar heating systems. I should note that entire books have been written on passive solar home design, so I can't do more than skim the surface of this subject here.

Passive Solar Design

You know how unbearably hot your car becomes when it is parked in the sun during the summer with its windows shut. This is passive solar heat run amok. Unlike your car, a properly designed passive solar home is sited and fashioned to absorb solar heat in the winter and to avoid it in the summer. The sun is the heat source, while the house and its windows, walls, and floors function as the collection, storage, control, and distribution system. The passive solar heating system operates for free and generally does not require electrical or mechanical devices to function.

Solar heat run amok.

When it is included in the original design of your home, a passive heating system doesn't materially add to construction expenses because you are simply making more intelligent use of the elements that would generally be in your house anyway. This makes passive solar the most cost-effective heating strategy, since once your house is constructed, the operating costs (except for a backup heating system) are basically zero. Under the right circumstances, passive solar elements also can be added or retrofitted into an existing home. There will be capital costs for the retrofit, of course, but afterward, the passive solar elements will act to reduce operating costs for heating your home.

Although passive solar design for buildings has been used for thousands of years in places such as China and Greece, simple passive solar principles are almost entirely absent from most modern homes in the United States. One of the main reasons why we don't make more intelligent use of these design elements is our current overreliance on fossil fuels. This has caused us to lose our sense of connection with the environment and has resulted in lazy and wasteful home design. We pay a price for dumb design in our houses by using far more energy than necessary.

Orientation

A solar home's primary design factor is its relationship to the sun. You can't expect to automatically harvest solar energy from any and every existing home. A home needs to sit on a suitable site for solar gain. Some sites just won't work, particularly those where tall evergreen trees, buildings, hills, or mountains obstruct the southern exposure. The best site for a passive solar home is one with unobstructed southern exposure coupled with protective slopes or large evergreen trees on the north side to block cold winter winds (in this case, the evergreen trees are an asset). Yet, even if you have a favorable site, an improper alignment of your house can make an otherwise good design fail. In general, a south-facing orientation that is within 30 degrees east or west of true south (the position of the sun at noon) will provide about 90 percent of the maximum solar collection potential. Southern orientation is crucial because the south side of a passive solar house is where the action is from a solar energy perspective.

The other key element of good passive solar design is to align various elements of your house, especially the roof, with the sun's best angle, which varies according to the seasons and the latitude of your home's location. The

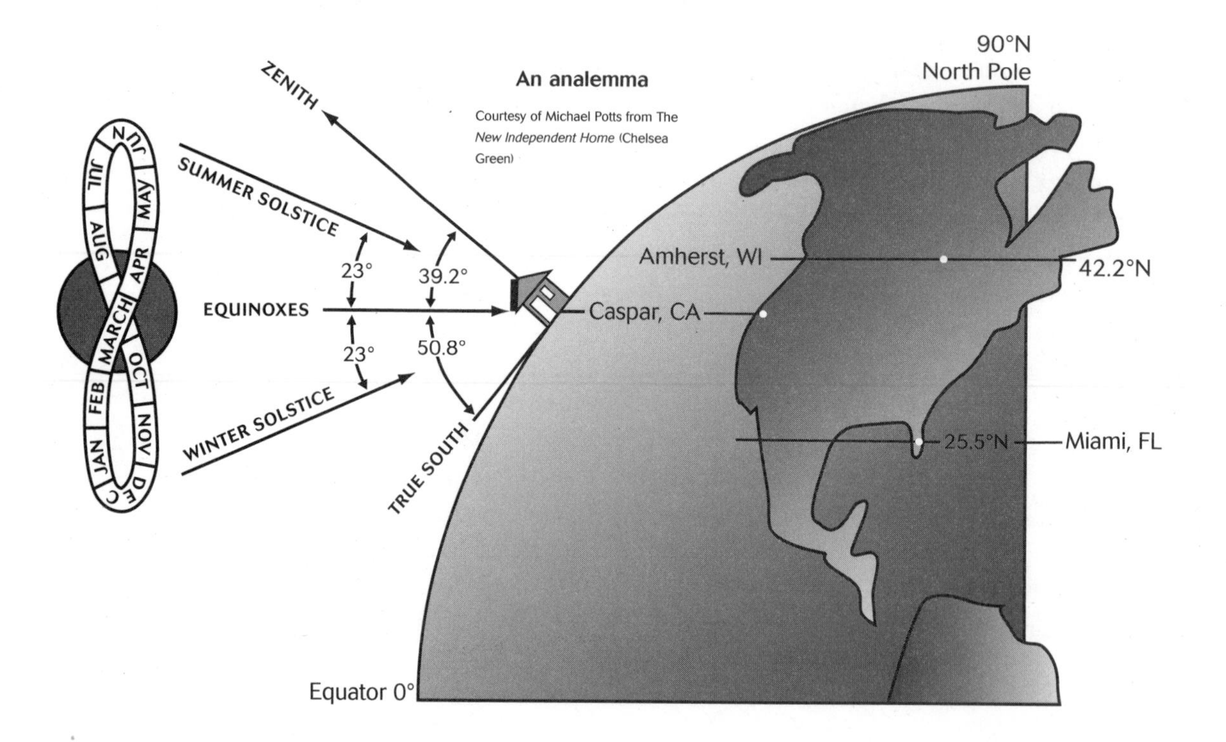

goal is to take best advantage of the sun at the times of the year you need it the most—generally in the winter. This is especially helpful if you are planning on installing roof-mounted active solar collectors (see chapter 6 for more details). An **analemma**—a graduated scale in the shape of a figure eight that represents the motion of the sun during the year—helps to demonstrate this phenomenon. This can get fairly technical, but the main thing to remember is that the sun is much lower in the sky during the winter and more nearly overhead in the summer (in the Northern Hemisphere). This can work to your advantage in passive solar design. If your roof overhang has been properly designed, the sun's rays will shine through your south-facing windows during the winter, providing free solar heat, while in the summer little or no direct sunlight will fall on those windows.

Heat Movement Basics

As I mentioned in Chapter 1, heat always moves from warmer to cooler materials until they reach equal temperature. A passive solar home makes use of this law of physics through three familiar heat movement mechanisms—conduction, convection, and radiation—to distribute heat throughout the living space. By design, heat is conducted efficiently through a variety of materials in a passive solar home. In addition, some passive solar homes use the convection of air to carry solar heat from a south wall into the building's interior.

Both solar radiation and infrared radiation are important in passive solar design. When radiation strikes an object, it is absorbed, reflected, or transmitted, depending on the properties of the object. Opaque objects absorb 40 to 95 percent of radiation from the sun, depending on their color; dark-colored objects generally absorb a higher percentage than light-colored objects. This is why solar absorbers usually have dark surfaces. The clear glass in south-facing windows of a passive solar home allows 80 to 90 percent of solar radiation to pass through. Once the interior surfaces of the home have absorbed this heat, the heat is radiated back from these surfaces as infrared radiation. The window glass reflects part of the infrared radiation back into the home, trapping much of the heat in the living space.

Design Techniques

A passive solar house contains the same materials as a conventional house, but the materials are arranged differently (more sensibly, in my opinion). The main difference between these two types of houses, then, is design. Designs for passive solar homes vary widely. Some passive solar homes are heated almost entirely by the sun. In others that simply rely on south-facing windows, the sun provides only a portion of the heat—this type of house is called a **sun-tempered house**. Even the first house that I built was sun-tempered. It just wasn't tempered very much.

Every passive solar home has five main elements.

- The **aperture** is the large glass window area that allows sunlight to enter a building.
- The **absorber** is the hard, dark, external surface of the heat-storage element.
- The **thermal mass** is material (such as water, brick, concrete, adobe, rocks) behind or below the absorber that stores the solar heat.
- **Distribution** is the method used to circulate solar heat from the collection and storage locations to other areas of the house. In a strictly passive solar house, distribution of heat is solely by conduction, convection, and radiation. Some designs also rely on mechanical devices such as fans and blowers.
- **Control** is elements that prevent summer overheating, such as roof and cantilever overhangs, awnings, blinds, or even landscaping and trellises. Other (more active) controls

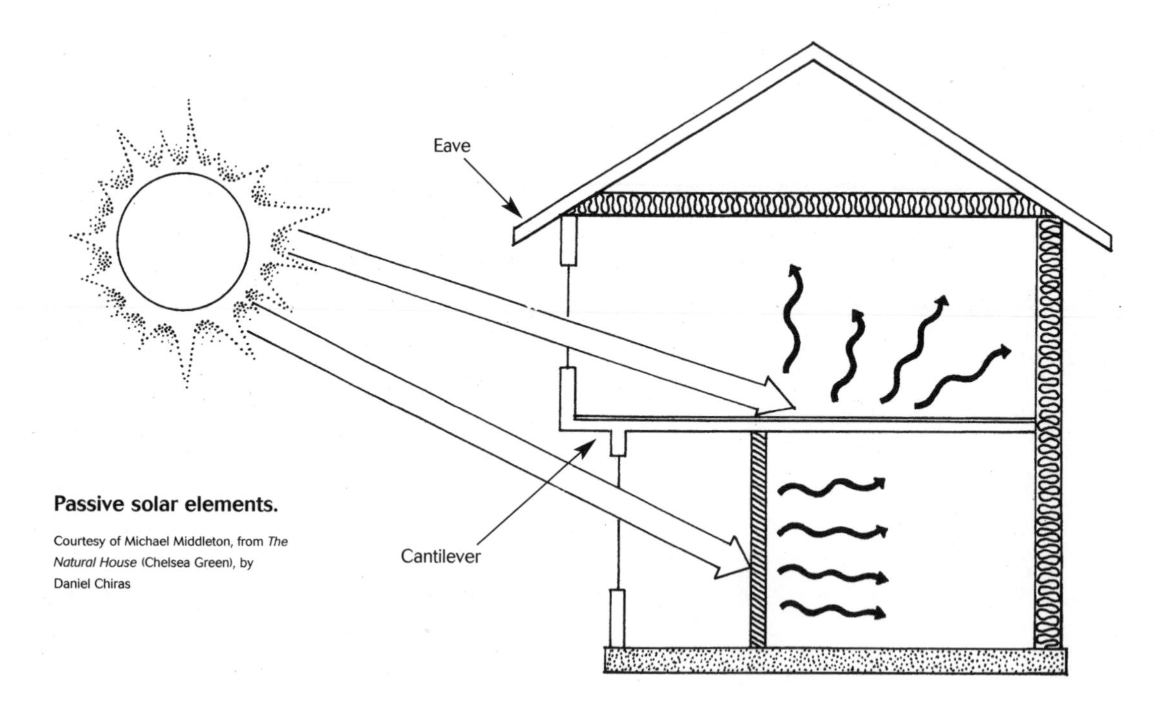

Passive solar elements.

Courtesy of Michael Middleton, from *The Natural House* (Chelsea Green), by Daniel Chiras might include thermostats for fans, vents, and other devices that assist or restrict heat flow.

Variety in passive solar design techniques results from considerations of how to combine these five design elements to best take advantage of the three main types of solar gain. **Solar gain** is the absorption of heat produced by the sun, and it can be direct, indirect, or isolated.

Direct Gain

Exploiting **direct gain** is the simplest passive solar design technique. In this strategy, sunlight passes through south-facing windows and then strikes dark-colored masonry floors or walls, which absorb and store the solar heat. After the sun sets, the room cools, and the heat stored in the thermal mass gently radiates and convects back into the living space. This is the strategy I used in the house I built in Lincoln.

In some passive solar homes, water-filled containers located in the living space are used to absorb and store solar heat. Water is an excellent heat sink because it can store twice as much heat as masonry per cubic foot of volume. There are several disadvantages to using water as a heat sink, however. Unlike masonry, water is not self-supporting; carefully designed structural support is needed beneath the heavy, water-filled containers. The water also has to be treated to prevent microbial growth.

The amount of passive solar heat that can be collected in a home depends on the area of south-facing windows, while the amount of heat that can be stored is determined by the amount of thermal mass in the design. It's important to balance the relationship between these two elements, because if the thermal mass is too small, the house (like your car in the summertime) will overheat. If the thermal mass is too large, it will still perform well, but after a certain point, the excess mass is just a waste of money. The ideal ratio between windows and thermal mass varies depending upon climate.

For thermal mass to be effective in a direct-gain strategy, the mass must be insulated from the outside (including the ground); otherwise, the heat absorbed will drain away quickly into the outside environment. This is why I laid rigid insulation under the gravel heat sink in my Lincoln home.

Installing thermal shades or drapes can reduce heat loss at night or prevent excessive heat gain in the summer. Unfortunately, I didn't install shades in my house in Lincoln, and much of the heat that was gained during the day radiated back through the glass at night.

Indirect Gain

Although it uses the same materials and design principles as direct gain, an indirect gain strategy places the thermal mass between the sun and the living space to be heated. The most common technique for taking advantage of **indirect gain** is to install a Trombe wall. A Trombe wall (named after French inventor Felix Trombe) is constructed in a south-facing room of a passive solar house. The masonry wall is 8 to 16 inches thick. A single or double layer of glass is installed about one inch away from the wall on the exterior of the house. The sun shines through the glass and on the dark surface of the wall during the day. Solar heat is stored in the wall and conducted inward; eventually, the heat radiates into the interior living space.

Heat travels through a masonry wall at the rate of about one inch per hour. It takes eight to ten hours for solar heat to begin to radiate into the living space from an 8-inch-thick Trombe wall. A Trombe wall keeps the living space at a comfortable temperature during the day and provides slow, steady heat for many hours after the sun has set. Roof overhangs or awnings protect Trombe walls from overheating during the summer. Although it is possible to install them in a retrofit project, Trombe walls are generally best suited for new construction.

Isolated Gain

The third type of solar gain used in some passive solar designs is **isolated gain**, which is solar gain in a specific space such as a sunspace, solar room, or solarium. Including a sunspace is a versatile strategy for solar heating that is appropriate for either new construction or an existing home. This makes the sunspace especially attractive for people who don't live in solar-designed homes. For existing homes, the best place to add a sunspace is on the south side of the house. If this is not practical, the sunspace can be located elsewhere as long as it has an unobstructed southern exposure and is directly connected to the house.

The simplest approach is to design a sunspace with a large area of south-facing vertical glazing and a thermal mass for heat storage. A masonry floor or wall(s) can serve as the thermal mass, as can water housed in tanks, tubes, or other containers. Heat distribution to the living space in the rest of the house is through windows, doors, ceiling- or floor-level vents, or sometimes fans. Sunspaces often undergo large swings of heat gain and heat loss through their large expanses of window glazing. Some of these extremes can be mitigated through the use of thermal mass and also with the use of low-e windows.

Installing doors or windows between the sunspace and the rest of your home gives you more flexibility in controlling heat gain (or loss) between the sunspace and your home.

Sunspace or Greenhouse?

It's important to understand that there is a distinct difference between a sunspace and a greenhouse. A sunspace is designed to provide heat and a pleasant room for your home. A greenhouse is designed to grow plants. The two design goals are not generally compatible. Greenhouses are usually designed with overhead or sloping glazing, which is fine for growing plants but not for creating a comfortable living space for people. The high humidity, molds, mildew, insects, dirt, and dust associated with greenhouses are also not welcome additions to most homes. What's more, it is difficult to shade overhead glass in a greenhouse during the summer, but relatively easy to shade the vertical glass in a sunspace.

Backup

Passive solar heat is free, renewable, and environmentally benign, and that sounds too good to be true. There *is* one major drawback to passive solar heating. No matter how well insulated your house is, it won't stay warm during those days (or even weeks) when the sun doesn't shine. While the sun is always shining out in space, the sun's vast energy potential doesn't always reach the surface of the Earth, particularly in regions where heavy cloud cover, fog, rain, or snow are persistent at certain times of the year, such as the Pacific Northwest and northern New England. Few heat sinks can overcome that kind of handicap. If you live in a passive solar house, you probably need a backup heating strategy.

I have often heard people dismiss passive solar simply because backup heat is sometimes needed. That's unfortunate, because, having a backup heating system is almost always wise, especially if you live in a chilly northern climate. Since most passive solar homes tend to be tightly constructed and well insulated, the backup heat source that helps you through a protracted sunless period can be fairly modest. And it can be renewable. A small woodstove, pellet stove, or masonry heater will do the job. Should you ignore the benefits of heating your home for free 80 percent of the time just because you may have to rely on a backup system 20 percent of the time? I don't think so.

Other Passive Strategies

If you live in a conventional home, how can you make better use of free solar heat? One very simple strategy is to open your window shades or blinds on sunny winter days to let as much sunshine in as possible. Remember to shut them again when the sun sets to reduce radiation of heat from your home at night. That's important, because windows account for between 10 and 50 percent of heat loss from your home. If your window shades are the insulating variety, all the better. We have gradually been replacing the old Venetian blinds in our home with insulated shades. This simple strategy has made a big difference. Our old boiler in the basement used to kick on repeatedly at night during cold weather. These days, it often does not come on at all during cold nights. That's a dramatic change.

Other techniques you can employ to create a partially passive solar home (or, at least, a sun-tempered home) are using water storage, adding insulation, installing south-facing windows, and replacing single-pane windows with high-performance windows.

Water Storage

If you have a room that receives lots of direct sunlight, you can reap solar gain by adding some thermal mass, such as specially designed tubes or tanks filled with water. Translucent fiberglass heat-storage tubes, for example, are available in a variety of diameters and heights. Filled with water, these tubes will store quite a lot of heat. You can dye the water a dark color so that it will absorb heat more quickly, choosing a shade to match your room décor. A line of tubes serves nicely as a room divider or privacy screen. In a pinch, even old 55-gallon drums painted black will work, but they're lacking in aesthetics—unless your interior décor is 20th-Century Industrial Warehouse.

If you add water storage tubes to a room, be sure that your floor system will support the added weight. Another retrofit option is a passively heated solar domestic-hot-water system (which I'll discuss in chapter 7).

Insulation

Appropriate quantities of correctly installed insulation are absolutely necessary for a passive solar home. Making sure that your home is properly insulated will pay off handsomely in lower heating bills. Often, an investment of a few hundred dollars in insulation will pay for itself in only two or three years.

Windows

No discussion of passive solar heating would be complete without windows, because windows play such a central role in any passive solar design. Installing additional south-facing windows can improve solar gain in any home and is a key passive solar retrofit strategy. Some people prefer to use plastic glazing materials instead of standard glass windows. Plastic can be cheaper, lighter, stronger, and easier to work with, and is especially popular for the do-it-yourselfer. However, plastics tend to scratch more easily than glass. Also, they expand and contract more and are harder to seal than glass. Plus, plastic glazing materials are a petroleum-based product.

Window technology has come a long way in recent years. If your home still has single-pane windows (almost half of the homes in the United States do) consider replacing them, if possible with windows that have an Energy Star label. Window glazing materials now come with a variety of special coatings and other features, and frames are available in aluminum, wood, vinyl, fiberglass, or various combinations of these materials.

New double- or triple-pane windows with high-performance glass are now available. In colder climates, select windows that are gas filled (argon, krypton, carbon dioxide, etc.) and feature low-e coatings on the glass to further reduce heat loss (for hot climates, spectrally selective coatings reject heat, but admit light). Low-e coatings improve the insulating value of windows by roughly the same amount as adding an additional pane of glass. While some low-e coatings lower total light transmittance, that reduction is more than offset by the increase in heat retained in your living space.

Almost 90 percent of windows currently manufactured are double-glazed, and roughly half have low-e coatings. The extra cost of these features only amounts to about 5 percent of the total window cost, so ordering windows with these features is highly advisable. Don't forget about the window frame, which represents about 25 percent of the area of most windows. Ask for window frames that are thermally nonconductive. High-tech windows in a passive solar home are very important, but they are also a sound retrofit choice for almost any home that has drafty old windows. High-tech windows are not cheap, but cheap windows can be the more expensive choice in the long run when you consider your home heating expenses over the life of your home.

Still More Options

If you'd like to learn more about passive solar, check out *The Solar House: Passive Heating and Cooling* by Dan Chiras (Chelsea Green Publishing, 2002) and *The Passive Solar House: Using Solar Design to Heat and Cool Your Home* by James Kachadorian (Chelsea Green Publishing, 1997). For additional information on passive solar homes, refer to the "Organizations and On-line Resources" section.

CHAPTER 6

Active Solar

To convert a conventional house to passive solar heating can mean a major, expensive renovation project. Sometimes, due to structural or other reasons, a passive retrofit won't work. Happily, active solar heating systems really shine in retrofitting strategies. In an active system, the sun is still the heat source, but a group of specially designed collectors harvests the sun's energy, which is then pumped or blown through pipes or ducts to your living space. Like their passive cousins, most active solar heating systems also have some kind of storage capacity to provide heat when the sun is not shining.

Active systems are divided into two main categories: liquid and hot air. Liquid systems are further subdivided into two categories: domestic hot water and space heating. Some active liquid systems actually provide both of these functions (more on this in chapter 7). Active systems can be expensive to install and require electricity to operate. An active system probably will not provide for all your heating needs, especially in cold, cloudy northern climates (a backup heater is necessary in these areas). The typical active solar heating system is not designed to meet 100 percent of your home-heating or domestic-hot-water needs. That's because this would not be cost-effective in most cases. Rather, active solar is designed to work in combination with other heating systems, which offers a lot of flexibility. As long as your expectations for an active solar heating system are realistic, you will probably be happy with the results.

History of Active Solar

Active solar heating systems have had a checkered history in the United States. Although there was an early boom in rooftop solar DHW heaters in

California and later in Florida in the early 1900s, the use of cheap and abundant fossil fuels killed the fledgling industry. The modern solar industry was founded in 1974, following the oil embargo of the Organization of Petroleum Exporting Countries (OPEC) of the previous year. Federal and state tax credits of the early 1980s for renewable energy systems gave the solar industry a major, if short-lived, jump start. Almost overnight, companies sprang up all across the nation to serve a seemingly insatiable demand for solar installations. President Jimmy Carter even had a solar hot-water heater installed on the roof of the White House. Then, the Reagan Administration pulled the plug on the incentives (significantly, the hot-water heater was removed from the White House). The fall of the industry was as sudden—and spectacular—as its rise. And the damage was severe and long lasting.

"Nine out of ten solar businesses went belly-up, and most of them didn't come back," recalls Leigh Seddon, the president of Solar Works, Inc. of Montpelier, Vermont. "That's interesting because solar heating is very cost-effective for hot-water heating, but the market has continued to shrink every year since 1985." Seddon, who founded his company in 1980, maintains that the solar heating industry has recently leveled off and may actually be growing slightly, but most of that activity is limited to areas where local utilities are running consumer-education and installation-subsidy programs. "So, in isolated spots there is growth," Seddon says, "and I think that at some point we will see the solar hot-water industry come back, but only when there is a level playing field in the energy industry." (The one part of the solar industry that has been growing dramatically in recent years is photovoltaic generation of electricity.)

Unfortunately, the solar industry has also had to struggle to overcome persistent public perception that it is too futuristic, unreliable, or impractical. None of this is true today, but there *were* problems with some faulty solar hot-water systems—and with some unscrupulous installers—during the solar boom of the late 1970s and early 1980s. Happily, the industry has matured, the technology has improved, and now there are performance standards that cover most components of active solar heating systems, making comparisons between products possible.

Before you decide to buy an active solar heating system, be sure to check local zoning ordinances, land covenants, and any other possible local restrictions that might cause you a problem. Homeowner-association rules, in particular, can be a real headache when it comes to solar collectors. These technicalities are not widespread, but in places where they are on the books, they can be hard to dislodge. If you run into a problem with a local restriction, you may need to spend time and energy to try to have the rule or regulation changed. Most of these restrictions are absurd and deserve to be changed, so it's probably worth the effort if you are committed to solar heating.

Active Solar Design

There are a number of basic design requirements for active solar heating systems that are similar to those for a passive solar house, but with active solar you'll have more flexibility in meeting those requirements. Ideally, your home should have an unobstructed southern exposure to the sun between 9 A.M. and 3 P.M. However, even if the south side of your house isn't oriented within 30 degrees of true south, you can place the solar collectors on your roof, walls, or even on the ground mounted on a framework at an optimum angle to take maximum advantage of the sun's rays. This flexibility in collector location can make up for a multitude of architectural and site problems.

A solar hot-water collector array mounted at ground level provides maximum location flexibility.

Courtesy of Solar Works, Inc.

Climate plays a major role in the design and cost of active solar installations. Active solar space-heating systems are a good choice in areas where utility rates are high and in climates that have long heating seasons during which there is a high proportion of sunny days. Active solar heating systems make less financial sense in areas with short heating seasons, frequently cloudy weather, or low utility rates and low fuel prices. This doesn't mean that you can't install an active system in the Pacific Northwest, northern New England, or Florida, but it does mean that, in those areas, cost-effectiveness would not be your primary motivation. The fact that an active solar home heating system is more sustainable and reduces the amount of pollution and greenhouse gases emitted from a home may be sufficient justification for some people to install one.

Sizing the System

An important factor in sizing an active solar heating system is the concept of **solar insolation,** which is a measure of the amount of sunlight reaching the Earth's surface. Solar insolation varies from place to place due to atmospheric conditions as well as the Earth's variable distance from the sun throughout the year. The unit of measure of solar insolation is called a **peak sun hour**.

Data on peak sun hours is readily available for many locations. Denver, Colorado, for example, has a yearly average of 5.7 peak sun hours per day, while Burlington, Vermont, has only 3.5. Thus, in order to capture the same amount of solar heat in Burlington as in Denver, a system would theoretically require a 60 percent larger solar collector array (an array is a group of collectors). Thus, for the same output, a system in Colorado will cost significantly less than a system in Vermont. I hasten to add that this does not mean solar-powered systems are not practical in cloudy locations; they just require a larger initial investment than in parts of the country that are blessed with more sunshine. Once installed, these systems will reduce your home's overall operating costs regardless of what part of the country you live in.

Active solar heating systems are normally designed to meet 40 to 80 percent of your home's heating needs. A system that provides less than 40 percent is generally not going to be cost-effective, unless it is a small installation designed to heat one room. The size of an active system helps to determine how much heat it can provide. Calculating the optimum size of a solar collector array is crucial, because this determines the size of the other components in an active solar space-heating system. Generally speaking, the solar collector area should equal 12 to 30 percent of your home's floor area. There are several ways of calculating system size more precisely, and you can expect your system installer to help you do so. Computer software programs are also available to help determine system size and other design aspects of solar heating systems (see the Organizations and On-line Resources section for information on sources of this software). In all cases, your house should be weatherized and insulated to high standards so that the active solar heating system can be as cost-effective as possible.

Collectors

The collector is the heart of most active solar heating systems. Solar collectors heat either air or a liquid. There are three main types of collectors for residential use: flat plate, evacuated tube, and concentrating collectors.

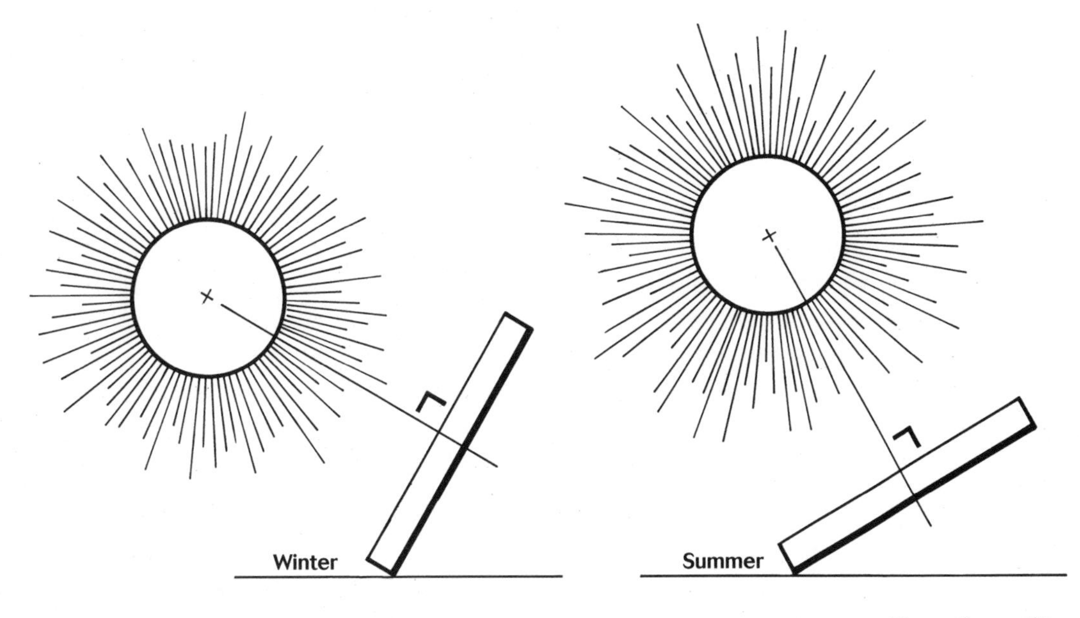

The optimum tilt angle in the summer and winter for solar collectors.

Under ideal conditions, solar collectors are as much as 90 percent efficient at converting solar energy to heat. The angle at which the sun's rays strike a collector is known as the **angle of incidence**. Because the sun is located at different heights in the sky at different times of the year, it would be ideal to adjust the angle of the collector (with respect to the plane of the Earth) as the seasons change, in order to keep the angle of incidence as efficient as possible for collecting solar heat. However, adjusting the collectors isn't always possible, so a compromise angle is often set.

For solar collectors that are mounted at a fixed angle, the general rule of thumb for optimum winter performance is to use the angle of latitude plus 15 degrees. The optimum angle in the summer is the angle of latitude minus 15 degrees. If your collectors are going to be set at a fixed angle, the typical strategy for hot-water collectors—using the winter angle—is probably the better approach because the heat lost during the summer will be negligible. This angle will also help to prevent summer overheating, and snow will slide off more easily during the winter.

When you are shopping for solar collectors, be sure to compare them for performance. Look for a Solar Rating and Certification Corporation (SRCC) sticker (or perhaps a state-issued certification sticker) on the equipment.

Flat Plate

A **flat-plate collector** is the most common type for home water- and space-heating systems. The typical flat-plate collector is an insulated metal box with a dark-colored absorber plate covered by glass or plastic glazing. The collectors may contain liquid or air. Sunlight passes through the glazing and strikes the absorber plate, which heats up the air or liquid inside the collector.

Installation Pointers

I want to mention a few quick pointers about installing roof-mounted solar collectors. In the case of liquid collectors, which can be heavy when they are filled with liquid, it is important to be sure that your roof structure is capable of handling the additional weight as well as the wind loading that the collectors may place on your roof. In addition, if your present roofing material needs to be replaced, it makes sense to replace it before you install the collectors, rather than afterward. Regardless of their condition, slate or clay roofing tiles present particular challenges to mounting solar collectors.

Evacuated Tube

Evacuated-tube collectors heat water for home heating applications that require higher temperatures (such as hydronic baseboard heaters). In this type of collector, sunlight passes through an outer glass tube, strikes an absorber tube, and changes to heat. The heat is then transferred to liquid flowing through the absorber tube. These collectors are named based on how they are manufactured: air is evacuated from the space between the tubes, forming a vacuum, which improves performance. Evacuated-tube collectors achieve higher temperatures and efficiencies than flat-plate collectors, but evacuated-tube collectors also cost more.

Concentrating

Concentrating collectors use mirrored surfaces to concentrate solar energy on an absorber called a receiver. A heat-transfer fluid flowing through the receiver absorbs the heat. Concentrating collectors can achieve high temperatures but only in direct sunlight; they perform poorly on cloudy or hazy days. Because of this, concentrating collectors are most effective in the desert Southwest.

Types of Active Systems

Because there are so many choices among active solar heating systems, it's a good idea to learn about all the possibilities and their relative advantages and disadvantages before you make a decision about using active solar. As I mentioned above, the two main categories of active solar systems are liquid and hot air. This refers to the heat-transfer medium used between the solar collectors, the storage tank or bin, and your living space. Liquid systems are more popular than hot-air systems because they generally cost less to operate, take up less space, and are easier to install as a retrofit. Hot-air solar heating systems are used mainly for space heating. Hot-air systems are not as popular as they once were due to problems with mold and mildew growth in some storage bins. Despite minor variations in design, most hot-air systems generally follow the same basic strategy and do not involve as many subcategories as liquid systems. We'll start with hot-air systems in this chapter and cover the many types of liquid systems in chapter 7.

Hot-Air Systems

An active solar hot-air system uses air as the medium for collecting and distributing solar heat for your home. Although there can be variations, the main components of this type of system are collectors, a rock-filled storage bin, fans, ductwork, and controls.

In a hot-air system, a fan draws air from your home through ducts to a series of channels in the space behind the dark absorber surface of the collectors, where the air is heated. The hot air then circulates into a storage bin, where it heats the rocks, which act as a heat sink. When the controls sense that the house needs heat, the warm air from the bin is blown to the living space through a ductwork system. When the temperature of the bin drops too low to supply enough heat to the living space, a backup heater kicks in.

Air systems have three operating modes, which I will call "primary," "secondary," and "third" modes. In the primary mode, the hot air moves directly from the collectors to your

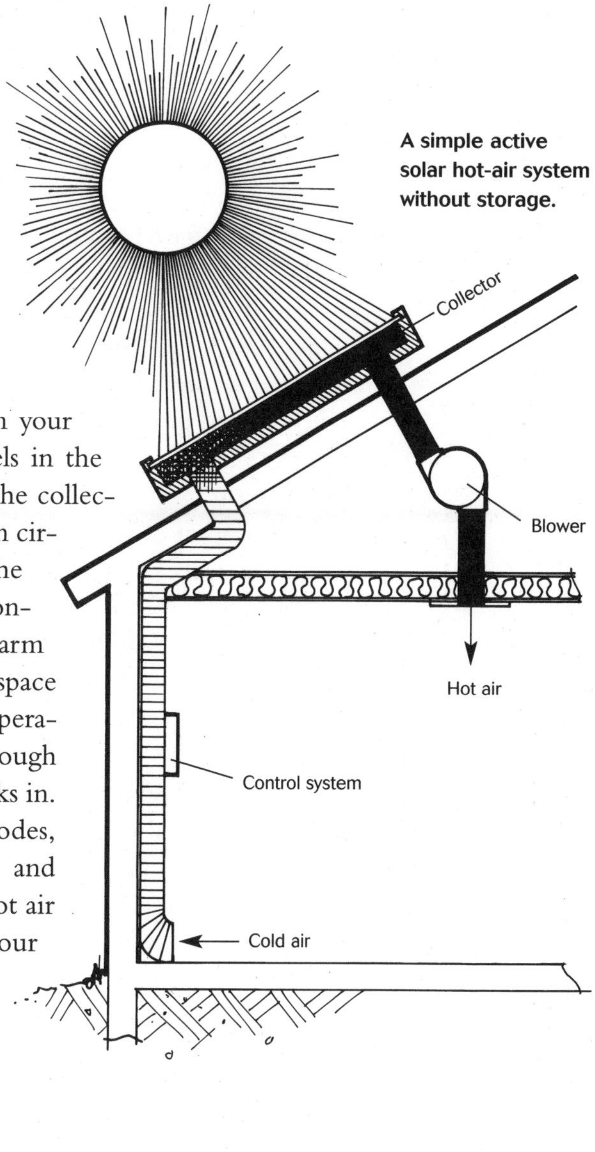

A simple active solar hot-air system without storage.

home's living space. In the secondary mode, excess heat is used to warm the rocks in the storage bin. In the third mode, your house is heated either from the bin or by a backup heater.

Storage

The storage bin usually has one or more plenums that control where the air is directed. The top of a rock-filled storage bin usually has a temperature close to 140 degrees Fahrenheit, while the temperature at the bottom of the bin may be about 70 degrees Fahrenheit. Storage bins can be made out of cement block, poured concrete, or wood, and may be placed in the basement, outdoors, or even underground. Underground bins must be thoroughly waterproofed, but even a bin inside your house should be sealed to prevent air or moisture leaks, which can seriously compromise performance. Storage bins should be heavily insulated. A rock bin placed in the basement or crawl space directly under the living areas of your home is the most efficient location for a storage bin because of the propensity of warm air to rise.

The rule of thumb for a rock bin is that it should be sized to provide between ½ and 1 cubic foot of storage for every square foot of collector, and should be 5 to 7 feet deep. Since rocks are not as efficient at storing heat as water, rock bins should provide two-and-a-half to three times more storage space than do tanks for a liquid system. The optimum-size rocks for the bin are ¾ to 1½ inches in diameter. The rocks should be thoroughly washed and dried before being placed in the bin. Once in the bin, the rocks must be kept dry to protect against the growth of mold and mildew.

Distribution

The distribution of warm air in a solar hot-air system is through ductwork. Ideally, this ductwork should be larger than that used by a conventional furnace because the temperature of the solar-heated air is cooler than air from a furnace. Thus, a larger volume of air is required to compensate for the lower temperature. The ductwork should be insulated to reduce heat loss.

Controls

The controls for an active solar space-heating system can be more complicated than those for a regular heating system. This is because the controls may have to analyze more signals and regulate more devices (possibly including your regular heating system). The controls for an active solar installation use a variety of sensors, switches, and/or motors to operate the different elements of the system and to provide for backup heat when necessary. Other controls may be used to prevent system overheating. At the heart of these controls is a differential thermostat, which measures the difference in temperature between the heat-storage bin and the collectors. When the collectors are warmer than the storage unit, the thermostat turns on the fan to circulate air through the collector to heat the storage unit in your house. Your system installer will help you to match the controls to your particular heating system.

Solar Room Systems

Not everyone wants (or is able) to invest in a large active solar space-heating system. If you're looking to limit your investment, a solar room heater offers a less expensive alternative that still involves the use of renewable energy. A solar room-heating system, as its name implies, is not intended to heat your entire home, and generally does not include any provision for heat storage. The basis for this system is a wall heater (often a group of them). A **wall heater** is a roughly three-by-six-foot solar collector attached to the exterior surface of a south-facing wall. The collector consists of a sealed frame (often made of wood) that contains a black metal heat-absorber plate behind two layers of glass. The sun shines through the glass, heats up the absorber, which warms the air inside the frame. The hot air rises and enters the house through an opening cut through the wall at the top of the collector. If a group of wall heaters is used, the air is often circulated through the collectors with the assistance of a thermostatically controlled fan that draws cooler air from the room into the bottom of the collectors through a second hole in the wall. When the sun is not shining, the thermostat turns off the fan, and dampers seal the openings in the wall to limit heat loss.

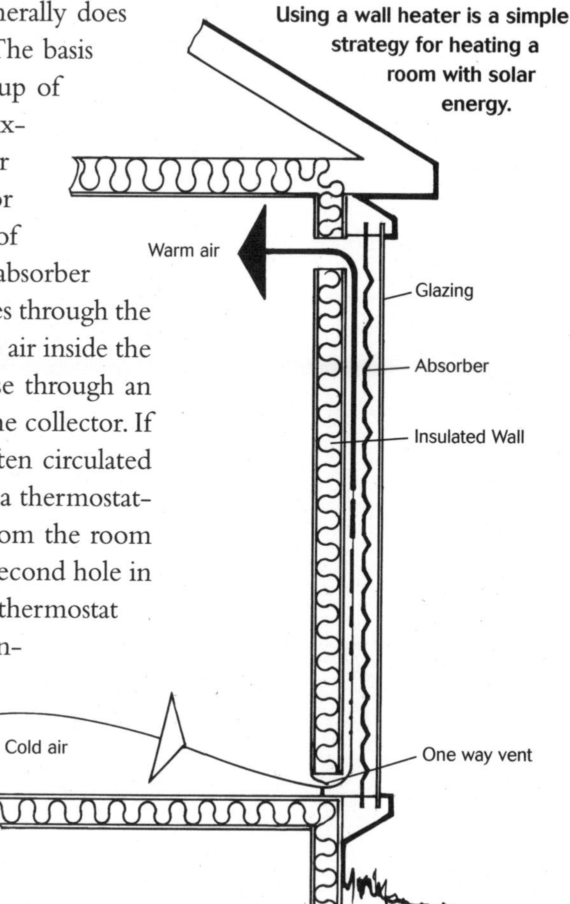

Using a wall heater is a simple strategy for heating a room with solar energy.

A wall heater is a fairly flexible device that can be sized according to the available space and is capable of providing substantial heat. If your roof overhang is properly designed, it will shade a wall heater in the summer. If not, you can install an awning or build a trellis to provide summer shade. A wall heater is relatively inexpensive to install, requires little maintenance, and will last for many years. A drawback is that a wall heater occupies what would otherwise be valuable south-facing window space.

Other Systems

There are other variations on hot-air space heating that do not rely on storage and simply have a series of standard hot-air collectors connected to a ductwork system. When the sun is shining, a thermostat turns on a fan that circulates room air through the collectors and back to the living space. There are even some siding and roofing materials that are designed to act as hot-air collectors. These materials can be connected to your home heating system via ducts and fans. The lack of storage capacity, however, limits the effectiveness of these strategies for residential purposes. On the positive side, solar hot-air heaters that directly heat the living space of your home without resorting to a heat sink can easily complement any existing heating system.

CHAPTER 7

Active Liquid Solar Heating Systems

When you investigate active liquid solar heating systems, you'll discover that you have many options for creating different kinds of systems that will match your particular climate, site, house design, needs, and budget. Liquid systems are especially well suited for retrofit projects but can also be included in new construction.

How Liquid Systems Work

Liquid systems are subdivided into two main categories: direct and indirect. Direct systems use water as the heat-transfer fluid in their collectors and pipes, while indirect systems use antifreeze solutions or a phase-change liquid (such as methyl alcohol) in the collectors and outdoor piping. Although there can be variations, the components in most liquid systems include hydronic collectors, a storage tank, pumps, one or more heat exchangers, and controls.

In liquid systems, the sun heats the liquid in the hydronic collector(s), and the liquid then circulates through pipes into your home. In a direct liquid system, the hot water from the collectors is circulated directly into the hot-water storage tank, from which it can be pumped through your home's hydronic heat distribution system. In an indirect system, the hot antifreeze solution from the collector is circulated through a heat exchanger in (or attached to) the hot-water storage

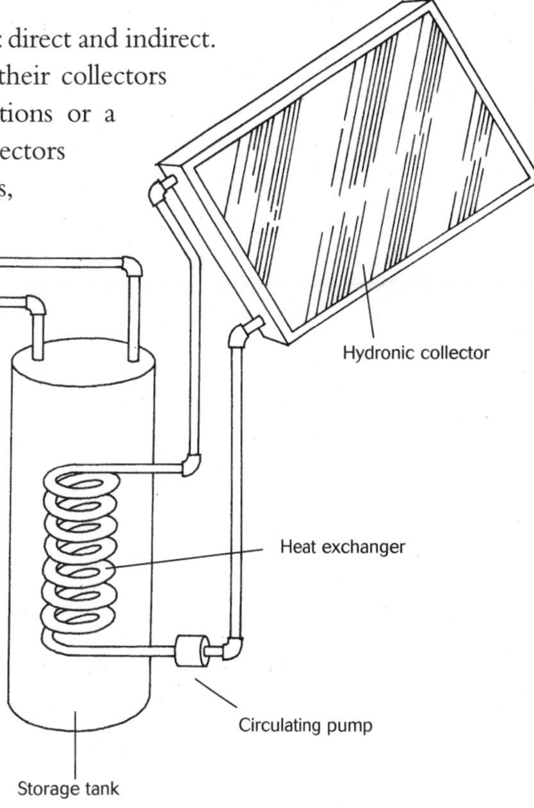

A liquid active solar heating system.

Courtesy of Michael MIddleton

tank, where it transfers its heat to the water in the tank. Then, the hot water in the tank can be circulated through your home's heat distribution system. Whether it's a direct or indirect system, the liquid stored in the tank and circulated through your home's hydronic heat emitters is almost always water.

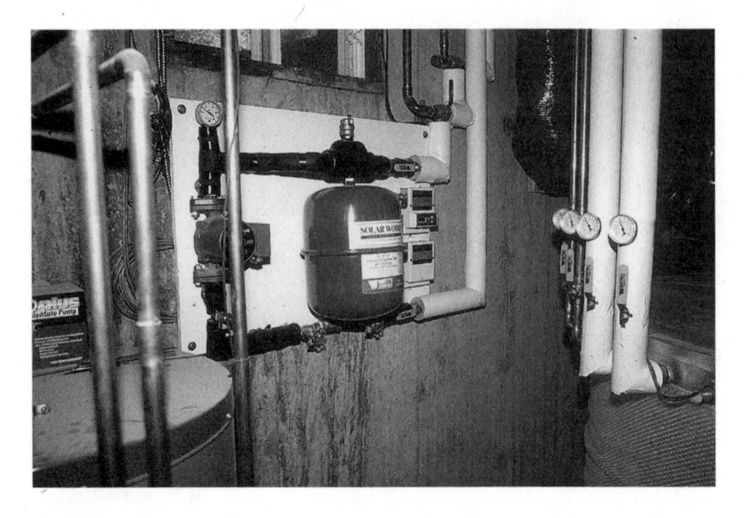

An active solar heating system pump/controls board.

Photo courtesy of Solar Works, Inc.

Liquid systems have three operating modes: primary, secondary, and backup. In the primary mode, the collectors send the heated liquid to a storage tank, while the tank also sends heated water to warm the living space in your home. In secondary mode, excess heat is stored in the storage tank but not circulated in the living space. In backup mode, if the solar heating system cannot supply enough heat (as on a cloudy day), it activates the backup system.

Storage

Liquid systems store their solar heat in water tanks or in the masonry mass of a radiant floor system. As a rule of thumb, most storage tanks need about 1½ gallons capacity for each square foot of collector area. The storage tank can be made out of steel, concrete, fiberglass-reinforced plastic, or wood. All types of tanks have temperature and pressure limits, must meet local plumbing or building codes, and should be insulated to reduce heat loss. Steel tanks are often used in new construction, but can be problematic in some retrofit situations because they may not fit through doorways. Concrete tanks are relatively inexpensive and can be poured in place, which makes them suitable for either new construction or retrofits. However, concrete tanks are heavy (even before they are filled with water) and normally need to be installed in a basement. Concrete tanks also need to be lined to prevent leaks, and connecting pipes to the tank can be problematic. Fiberglass-reinforced plastic tanks do not corrode, but special care must be taken to ensure that the temperature of the water does not exceed the manufacturer's specifications. Wooden tanks are relatively inexpensive and can be used for new or retrofit situations; these tanks must have a plastic liner.

Distribution

A distribution system for liquid systems has two main sections. The first is the piping and pump that circulates the liquid from the collectors to the main heat exchanger and storage tank(s). The second segment includes piping and one or more pumps that distribute the hydronic heat from the storage tank to standard heat emitters in the living space, such as radiant floors, baseboards, or radiators.

If you already have hydronic radiant-slab heating in your home, adding a liquid solar heating system could be an excellent strategy. Radiant-slab heating is the most compatible with liquid solar heating systems because, in addition to storing heat, the slab also performs well at the lower water temperatures (90 to 120 degrees Fahrenheit) typical of a solar heating system. Baseboard hot-water heating elements and radiators are not as good a match with solar heat because they typically require water temperature between 160 and 180 degrees Fahrenheit (evacuated-tube or concentrating collectors can help in this situation). Whatever type of heat emitters you use, the existing boiler connected to them can serve as your backup heater for times when the sun isn't shining.

If your home has an existing forced hot-air heating system, you can integrate it with a liquid solar heating system. A fairly common approach is to pump hot water from the storage tank of the liquid system to a heat exchanger in a standard hot-air distribution system. The solar heating part of the system is still considered to be "liquid," while the living space distribution system is obviously hot air. In this case, your existing hot-air furnace becomes the backup heater.

Solar Domestic-Hot-Water Systems

A solar domestic-hot-water system (SDHW) heats only your domestic hot water, while a solar space-heating system is connected to, or integrated with, your home's heating system. Active SDHW systems are often part of a passive solar house design, while active space-heating systems usually are not, since a passive solar home normally provides its own space heat. Combining space heating and domestic-hot-water systems can result in greater efficiency and fuel savings, as I'll explain later in this chapter.

Domestic-hot-water heating is important because it is usually the second-largest utility expense in the average home (space heating is the largest).

Recirculating

In a direct system that employs a **recirculating** strategy, when the temperature in the collectors approaches the freezing point, the system pumps heated water from the storage tank back through the collectors to prevent freezing. The obvious disadvantage of this approach is that the temperature of the heated water is reduced by radiation of heat when the water passes back through the collectors, wasting much of the heat the water originally gained from the sun. Also, if the recirculating components of the system should fail for any reason, the collectors and exterior piping may freeze, potentially causing serious damage. For these reasons, the recirculating strategy is not as popular as other strategies.

Indirect Systems

Indirect solar hot-water systems, also known as closed-loop or antifreeze systems, use an antifreeze solution (a glycol-and-water mixture is the most common) or a phase-change liquid (such as methyl alcohol) to keep the collectors and exterior piping from freezing. The antifreeze or phase-change fluid transports the heat from the collectors to the heat storage tank, where a heat exchanger transfers the heat to the water in the storage tank. (This type of heat exchanger is a liquid-to-liquid heat exchanger.) Most of the problems associated with direct systems are eliminated in indirect systems.

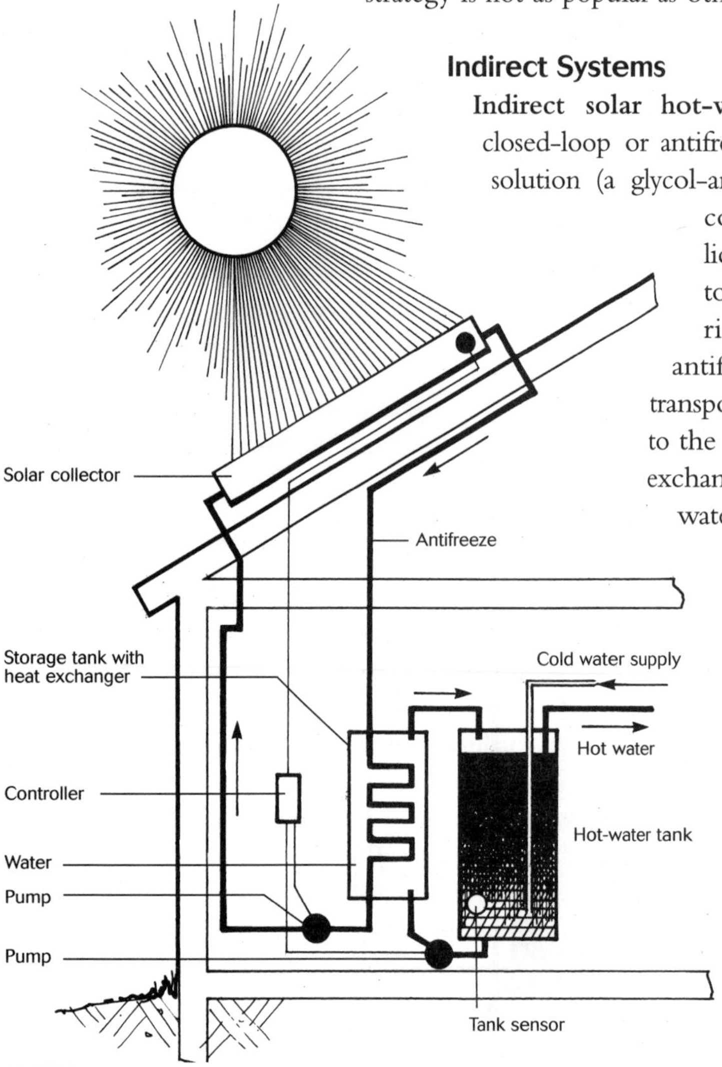

Solar collector
Antifreeze
Storage tank with heat exchanger
Cold water supply
Hot water
Controller
Hot-water tank
Water
Pump
Pump
Tank sensor

An indirect active solar hot-water system with two storage tanks.

An indirect system that heats domestic hot water should have a double-walled heat exchanger to prevent contamination of household water with antifreeze or other fluids. Some heat exchangers are mounted outside the storage tank; others can be located inside the tank. Some storage tanks have more than one heat exchanger, depending on the number of systems that are connected to the solar heating system; some systems have more than one storage tank.

Storage tanks for indirect liquid systems should be sized to hold 20 to 30 gallons of hot water per person. The storage tank (and, if practical, the collectors) should be located as close as possible to the household distribution and backup system to minimize heat loss from the pipes. Insulating the piping will also reduce heat loss. Since all the piping in an indirect liquid system is protected against freezing, it can be installed in any unheated location, which may offer additional design flexibility. Because antifreeze tends to degrade over time, however, systems that use antifreeze need regular inspections to ensure that the antifreeze solution is still effective.

In most indirect systems, a small photovoltaic (PV) module can be used to operate the circulating pump, creating a self-powered, self-controlled system. This type of installation runs when the weather is sunny and automatically shuts itself off when the sky is dark or cloudy. Whether powered by a PV module or some other source, indirect systems are the preferred option for extremely cold climates. My wife and I plan to install a PV-controlled indirect SDHW system on the south-facing garage roof of our home.

Thermosiphoning System

Another type of SDHW system is a **thermosiphoning system**. In this design, which can be either direct or indirect, the collector(s) is located below the heat-storage tank, often on the ground. The heated water (or antifreeze) flows by natural convection out of the top of the collector up to the storage tank (or, in the case of antifreeze, up into a heat exchanger), heating the water in the storage tank. Then, the cooling water (or antifreeze) flows back to the bottom of the collector, where the thermosiphoning circulation continues. Technically, this is a passive system, but it utilizes most of the same components as its active cousins, and so it seems appropriate to discuss it here. The beauty of this design is that it does not require any pumps to operate, but it is admittedly less efficient than an active system.

Since the storage tank needs to be at least 2 feet higher than the collector, the tank is often mounted on the second floor, in the attic, or even on

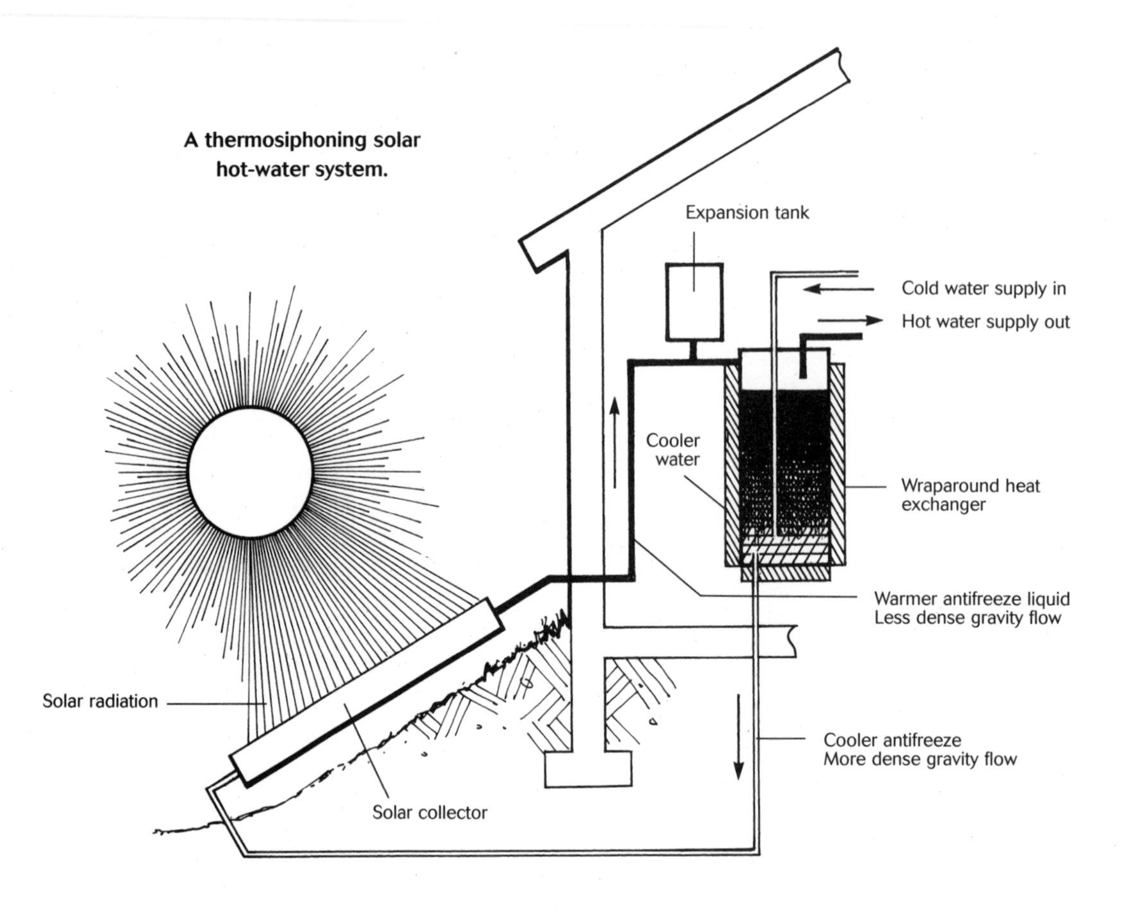

A thermosiphoning solar hot-water system.

the roof. In this scenario, you need to be certain that your house framing will support the weight of the water-filled tank, and that the water in the tank will not freeze in the attic or on the roof. Direct thermosiphoning systems are generally restricted to areas where temperatures do not fall below freezing during the winter, or perhaps where they would only be used in the summertime. In all other locations, an indirect system is called for. This type of system would not be practical in homes where a ground-mounted collector or elevated heater are not feasible. So thermosiphoning systems do have limitations, but they are relatively inexpensive to install and easy to maintain.

Batch Heater

A batch heater is the simplest direct system of all. It's a passive SDHW system and is best suited for relatively mild climates. A **batch heater** consists of

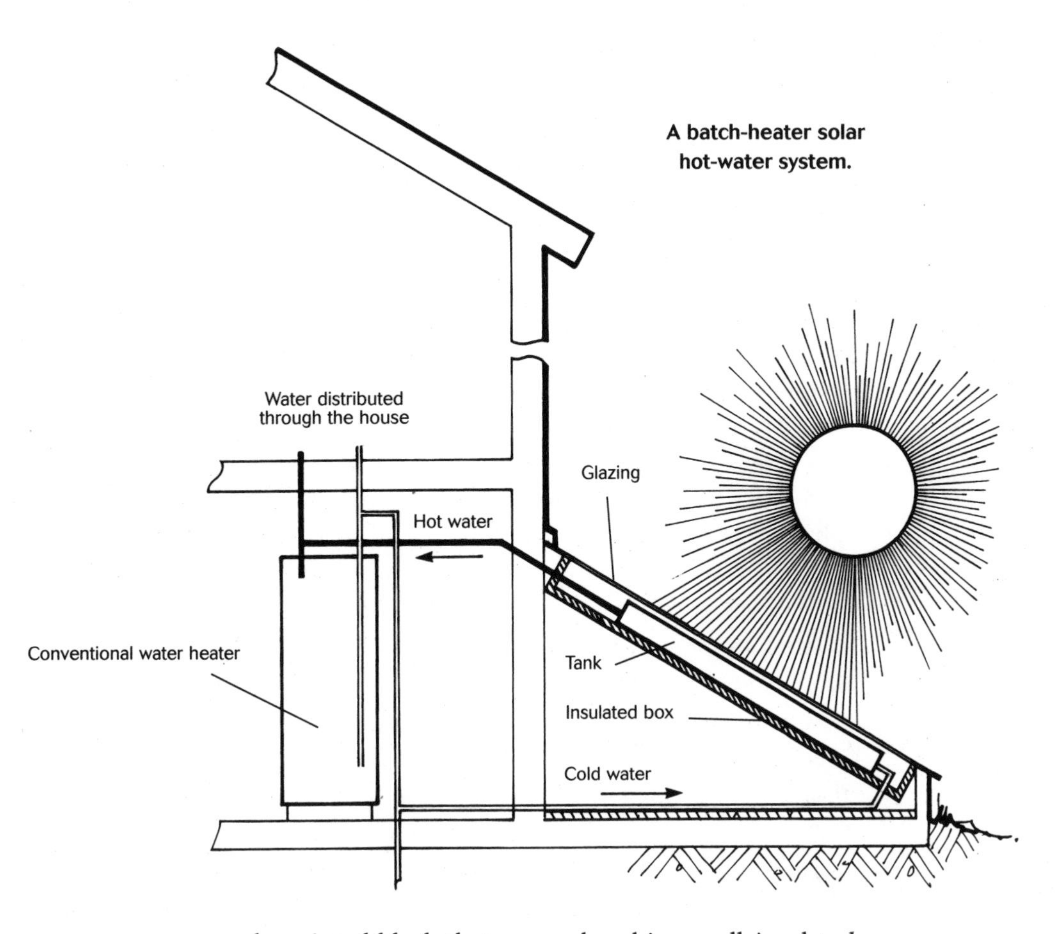

A batch-heater solar hot-water system.

one or more water tanks painted black that are enclosed in a well-insulated box covered with glazing. The sun shines through the glazing and heats the tank(s), which transfers the heat to the water inside. In this design, the collector and the storage tank are combined. The tank(s) is connected directly to your home's domestic-hot-water system. When you turn on a hot water tap in the house, hot water is withdrawn from the batch heater and is replaced by incoming cold water.

A batch heater should be placed as close to the domestic-hot-water system as possible to minimize the length of interconnecting piping. The pipes to and from the batch heater need to be heavily insulated. Some types of commercially manufactured batch heaters and piping are specifically designed to minimize or withstand freezing. A batch heater is inexpensive to install, requires almost no maintenance, and is very cost-effective.

Solar Space Heating

An active solar space-heating system uses the same components as you'll find in most types of active SDHW installations. However, in a space-heating design, the components are connected to the heat distribution system in your home. That distribution system can include hydronic radiators or radiant floor heat in a concrete slab, or even a forced hot-air system (with the use of a liquid-to-air heat exchanger). But because the heating requirements of a space-heating system are far greater than for an SDHW system, larger collection and storage capacity needs to be designed into the system. As a consequence, most liquid active space-heating systems include two heat-storage tanks (or one very large tank). The first tank contains the heat exchanger that transfers heat from the antifreeze solution to the hot water. That hot water often flows into a second tank that may contain additional heat exchangers (I'll explain why in a moment).

While using solar energy to heat residential living space is a viable strategy in most locations, it does have limitations. The most obvious is the impracticality of meeting 100 percent of your home's heating needs with an active system. In virtually all cases, you will need backup heat. In the solar industry, there are general guidelines about **solar fraction**—the portion of your heat that can be provided by a typical active solar heating system. Leigh Seddon of Solar Works, Inc. in Montpelier, Vermont, explains. "Achieving 100 percent solar space heating is incredibly costly, so we have some rules of thumb that we use for various regions," he says. "For the New England region, the rule of thumb is that about 60 percent solar fraction is the cost-effective stopping point. After that, you want to use a supplemental heat source to fill in the rest. Whereas, if you go to Albuquerque, New Mexico, you're talking about perhaps 85 to 90 percent solar fraction because the weather is so consistent and there is so much sun." Except for researchers or purists who are willing to pay the extra costs, almost no one tries to achieve a 100 percent solar fraction; 85 to 90 percent is the practical maximum, according to Seddon.

In the winter, a solar space-heating system just can't produce all the heat you need. Ironically, the other major problem with an active solar space-heating system is that it produces too much heat in the summer. Think about it; you've got all that hot summer sun striking the large solar array on your roof, producing hundreds of gallons of superhot water. But what can you do with it? You certainly don't want to heat your house in July. If you can't figure out a productive use for this hot water, then your only choice is to let your large, expensive solar heating system sit idle for much of the year. This is where a hybrid system comes in handy.

Hybrid Systems

A **hybrid system** combines two or more heating functions in the same system, typically space-heating and domestic hot water. The SDHW part of the system provides a place to use the solar-heated water year-round, which makes the investment in active solar equipment more cost-effective. Typically, a hybrid system includes a hot-water storage tank that contains more than one heat exchanger. The lowest heat exchanger in the tank is heated by the solar heating system, a middle heat exchanger can be supplied by a backup heating system, while a top heat exchanger can provide heat for your space-heating system. This arrangement provides more flexibility than a single-purpose solar heating system, but sometimes it's still not flexible enough. A large solar space-heating system often produces more hot water during the sunny summer months than can be utilized. In this scenario, your solar heating system is in danger of overheating. What to do?

The Pool Strategy

If you happen to have a swimming pool, you don't have to look far for an answer to the summertime overheating dilemma. Your pool is the perfect place to dump excess hot water from your active solar heating system. Not only will this strategy make your system even more cost-effective, but also it has the added benefits of eliminating the need for a pool heater and extending the length of your swimming season. If you don't have a swimming pool, you can pipe the water into a hot tub instead, although a hot tub does not provide as much heat-sink potential as a swimming pool. Obviously, if you don't already have a pool or hot tub and decide to add one, this strategy will increase the cost of your system. But a solar-heated hot tub certainly is a neat idea.

Maintenance

Like any mechanical heating system, an active solar heating system needs routine, periodic maintenance in order to function efficiently and safely.

There are some basic maintenance functions that you can perform yourself, while other system checks or repairs are best left to a professional. Tasks that you can probably do yourself include cleaning your collectors if they become dirty or excessively dusty, and trimming trees or bushes that have started to shade your collectors (if it's a very tall tree, call a professional).

If you are reasonably handy with basic household maintenance projects, you can also tackle the following maintenance tasks.

- Check for cracks in collector glazing and replace if necessary.
- Check your plumbing or ductwork for leaks and repair as needed.
- Check the insulation covering pipes, ducts, and wiring for wear or damage.
- Check flashings and seals around roof penetrations for leaks and caulk or seal if necessary.
- Check nuts and bolts on the collector support structure and tighten them as needed.
- Check pump(s) and blower(s) to be sure they are operating properly; if not, get them serviced.
- Check the storage tank(s) for leaks or signs of corrosion.

Virtually all other repairs or replacements of system components are best left to a professional. This includes checking the heat transfer fluids used in indirect liquid solar heating systems. These fluids need to be checked every few years, and will eventually need to be renewed or replaced. If your system uses a phase-change fluid, a professional *must* perform this maintenance work.

Backup

I have repeatedly stressed the important role played by a backup heater in any active solar space-heating strategy. I've implied that the backup system is connected to, or integrated with, the solar heating system. While this is an excellent strategy, particularly if you tend to be away from home during the day, it's not the only one. Depending on your home's design and whether someone is at home most of the time, you might find that a separate wood-, pellet-, or corn-fired stove can easily provide the backup heat you need for cloudy days (more on these stoves in parts 3 and 4).

Costs

How much does an active solar heating system cost? Typically, many active solar energy systems have high initial installation costs. However, once installed, these systems have extremely low operating costs. While a few of the smaller, simpler systems are relatively inexpensive, the majority of the more complex ones are not. The high initial cost can be a real challenge for many people (myself included), but the long-term advantages can be substantial. System costs vary widely depending on the size and type of installation and on whether the active solar heating system can be integrated easily into your existing heating system. A liquid active system, for example, might make a lot of sense from an installation and operational standpoint in a home with an existing hydronic radiant-floor heating system and a pool or hot tub, but less sense if you only have a forced hot-air system with small ductwork.

SDHW heating systems generally cost about $1,000 for a simple do-it-yourself system, and up to $4,000 for more sophisticated systems. Payback periods are four to ten years, depending on the cost of the fuel that is being replaced. A simple one-room solar wall heater may cost only about $800, but an active solar home heating system starts at about $5,000 and can exceed $18,000. Plus, there's the need for backup. If you already have a conventional heating system and are installing the solar heating system as an enhancement or retrofit, your conventional system will become your backup. Eventually, you will recover most or all of your capital investment in an active solar heating system through lower operating expenses—but that will take quite a few years. In the meantime, you will have the pleasure and satisfaction of heating your home comfortably and renewably with the sun—most of the time.

If you are seriously considering an active solar heating system for your home, I strongly advise you to seek competent advice from an experienced solar heating designer and have the installation work done by a professional. A well-designed, properly installed solar heating system will provide considerable long-term benefits, but a poorly designed or installed system will be a waste of your hard-earned money.

PART THREE

HEATING WITH WOOD

CHAPTER 8

Working with Firewood

Burning wood is more than just a heating strategy: it's a lifestyle that offers many pleasures, as well as its share of frustrations, potential dangers, and responsibilities. It might be helpful to think of your wood-fired heating appliance (woodstove, furnace, masonry heater, or fireplace) as a small child or pet. It depends on you, it requires commitment for the long term, and it's a lot of work. In return, it provides you with a great deal of comfort and pleasure. After all, there's nothing quite as satisfying as sitting by a cozy, hot fire on a cold winter's night during a power failure. When used as your primary source of heat, a wood-fired appliance can become a central part of your lifestyle and daily routine.

For many years, wood-burning appliances (freestanding woodstoves, fireplaces, and fireplace inserts) made up the vast majority of so-called "hearth appliances" in the United States. But that's changed. From the early 1990s to 2000, gas-fired hearth appliances made dramatic gains, and they now outnumber all wood-burning appliances combined. In 2000, according to the Hearth, Patio & Barbecue Association, 62 percent of hearth units were gas-fired, 36 percent used cordwood, while only 2 percent used pellet fuel. From these statistics, it's clear that in the past decade we have moved even farther away from sustainable home heating practices. If we are going to free ourselves from our overreliance on fossil fuels, we need to reverse this trend.

The responsible burning of wood and other biomass fuels is one way to help. While wood may not be the best choice of fuel in densely populated urban areas where air is already heavily polluted, burning wood does make sense in suburban, small-town, and rural areas. This section of the book details the many ways to burn wood safely in a wide variety of heating appliances; part 4 covers options for burning other biomass as well.

A Recent History

Although the oil embargo of 1973 generated plentiful sales for woodstove manufacturers, a much larger boom took place beginning in 1979, as a result of the Three Mile Island nuclear accident and the oil shocks associated with the taking of American hostages in Iran. "These incidents gave the woodstove industry an enormous shot in the arm," recalls Craig Issod, the Webmaster of HearthNet. "All of a sudden there were perhaps 450 manufacturers of airtight stoves in the country. Every welding-shop owner and his brother was in on it, and they were all selling stoves because demand exceeded supply. But after things settled down in a year or two, when people saw that oil wasn't going to be three dollars a gallon, it became obvious that many of these manufacturers were not going to last. In the next few years, vast numbers of them went out of business; out of the 450 manufacturers, perhaps 50 remained."

When Ronald Reagan became president in 1980, virtually all support for the renewable energy industry ended. "That was the end of any national focus on these issues," Issod says. "At about the same time, European coal stoves became increasingly popular in the United States. People started to say that no one was going to burn wood any more. Well, I've heard that many times over the years.

"In the late 1980s, a lot of people started saying that you should get rid of your woodstoves and burn gas instead," Issod continues. "Then, in 1991, the Gulf War caused another big oil price shock and wood became popular again. That was followed by a move to wood pellets, which were touted as an easy way to burn wood in places that didn't have access to firewood." Throughout this period, however, woodstoves continued to sell steadily. In the late 1990s, when the Y2K scare spurred fears of power supply system disruptions, woodstove sales soared again. Next, the California energy crisis and the price spike in natural gas resulted in a significant boost in pellet stove sales as well. Despite all the ups and downs over the years, wood has maintained a dominant position in the renewable fuel sector, according to Issod.

Where There's Smoke...

One issue has dogged the wood-burning appliance industry for years: air pollution. It was difficult for an industry that promoted itself as being environmentally beneficial to ignore the fact that woodstoves and fireplaces polluted the atmosphere. In the early 1980s, state and Federal officials began to monitor emissions from wood-burning appliances. In response, the industry began to develop plans for including a catalytic converter in woodstoves to reduce emissions. Initially, it seemed to be a good idea. From an engineering standpoint, it was a relatively simple matter to add a catalytic converter to most existing stove designs, and the converter did reduce particulate emissions. When the U.S. Environmental Protection Agency (EPA) announced that the sale of high-emission woodstoves would be banned after July 1, 1992, the number of stoves equipped with catalytic converters multiplied exponentially. Unfortunately, the customer response to the new catalytic stoves was generally not favorable, and customer complaints multiplied as well.

As the complaints mounted, many stove manufacturers began to shift from the catalytic design to an alternative, noncatalytic (or clean-burning) technology first developed in New Zealand. This clean-burning technology was adapted to meet EPA standards and was quickly introduced into stoves with plate-steel construction. Adapting cast-iron stoves to this technology was more difficult, however, because it involved time-consuming and expensive changes in design and manufacturing processes. The clean-burning, noncatalytic design has undergone further refinements and is now found in the vast majority of stoves on the market. One other result of the tighter emission standards for woodstoves was a dramatic decline in the number of companies making stoves. Today, there are perhaps ten major manufacturers that sell woodstoves in the United States, a far cry from the hundreds of prior years.

Types of Firewood

Anyone who is serious about burning wood in a wood-fired heating appliance needs to have a readily available supply of firewood. But what kind of wood makes the best fuel? The rules for choosing wood apply equally to almost all wood-burning appliances (except for a masonry heater, which requires smaller-diameter wood, as I'll explain in chapter 14). Firewood can be hardwood or softwood, but within those broad categories, there are dozens of species of trees. Which ones are best for firewood?

A general rule of thumb is to choose the heaviest wood available to you. Whatever species of tree it comes from, any given weight of wood contains about the same heat potential. A pound of pine (a softwood) and a pound of oak (a hardwood) will both produce about seven thousand Btu, but the pine takes up more space. This is important because wood is sold by volume, not weight. In addition, your heating appliance's firebox has a limited capacity in terms of how much wood it will hold; this means denser hardwoods are a better fuel because they have more heat potential for any given volume. You don't have to avoid burning softwood altogether. In some parts of the country, hardwood is simply not available, and softwoods such as fir, aspen, and pine are regularly burned for fuel. The main point is to use the best of whatever species are available in your area. To get an idea of what your yearly cost will be, you may want to try using the online calculator at www.hearth.com/fuelcalc/woodvalues.html, which calculates the annual cost of heating with various species of wood.

Fuel Values of Some Woods

High (20–27 million Btu per cord)	Medium (17.5–20 million Btu per cord)	Low (12.6–17.5 million Btu per cord)
American beech	Black cherry	Balsam poplar
Apple	Black gum	Basswood
Black birch	Black walnut	Black willow
Black locust	Elm	Box elder
Blue beech (American hornbeam)	Gray birch	Butternut (white walnut)
Crabapple	Holly	Catalpa
Dogwood	Honey locust	Chestnut
Eucalyptus	Magnolia	Cottonwood
Hickory	Oregon ash	Largetooth aspen
Hop hornbeam (hardhack, ironwood)	Red gum	Quaking aspen (poplar)
Live oak	Red maple (soft maple)	Red alder
Persimmon	Sassafras	Tulip poplar
Shadbush	Silver maple	
Sugar maple (hard maple)	Sycamore (buttonball, buttonwood)	
White ash	White birch	
White oak		
Yellow birch		

Sources of Firewood

Obtaining firewood requires an investment of your time. How much time depends on whether you decide to cut your own firewood or have someone else cut it for you. Even if you pay someone to cut, split, and deliver your firewood, you will still have to stack the wood and then handle it again when you burn it. Inescapably, burning wood involves some physical labor, as well as the need to clean up the inevitable mess that results when you bring firewood into your home.

Cutting Your Own

The key to ecologically sound wood-energy use is to ensure that your firewood is being harvested sustainably. If you have your own woodlot, or have access to one, you can ensure sustainable forestry practices by cutting your own firewood. If you are young, healthy, have a strong back (and a weak mind, as the old saying goes), enjoy outdoor exercise, and are competent to handle a chain saw, cutting firewood may be just what you have been missing in life. In my younger days, I spent many pleasant hours working in the woods to gather next year's firewood supply. It's great therapy and can be very satisfying work.

I don't want to overidealize the task of cutting firewood. Working in a woodlot is hard labor and can be extremely dangerous for the novice (or inattentive) woodcutter. For safety's sake, it's very important to have the right equipment. In Vermont, almost every year, someone is killed in the woods while cutting firewood. Often, the victim is an experienced logger.

On the positive side, why pay high fees to a health club to keep yourself in shape when you can do it almost for free and end up with next winter's fuel supply at the same time? However, if you dislike the idea of spending hours in the woods with a chain saw, you'll want to buy your wood instead.

A Good Supplier

Buying firewood is more complicated than purchasing most other fuels. That's because wood is a natural product that varies from tree to tree, from species to species, and from supplier to supplier. Firewood suppliers also tend to be an individualistic lot. If and when you find a reliable supplier who harvests wood sustainably, do all you can to maintain a good working relationship.

A good supplier of firewood can sometimes be hard to find. Courtesy of Jeff Danziger

The length of firewood you need is a key issue when you buy wood. One common mistake made by novices is to measure the firebox of their appliance and then order firewood just slightly shorter than that measure. I've learned firsthand that most firewood suppliers don't cut wood that accurately, and there is nothing more frustrating on a cold winter's night than to discover that a chunk of wood is half an inch too long to fit in your stove or furnace. Unless your supplier has a mechanized operation that produces accurate lengths, it's a good idea to deduct at least 4 inches from the firebox measure when you order your wood.

Figuring out exactly *how much* wood you are buying can be an interesting exercise. The basic measure of firewood is the cord (hence the term "cordwood"). A **cord** is a measure of the volume of a stack of wood that is 8 feet high, 4 feet long, and 4 feet deep. A full cord theoretically contains 128 cubic feet. As a practical matter, there are air spaces between individual pieces of wood when they are stacked, so the solid volume of wood in a cord can range from 65 to 100 cubic feet. To complicate things even more, wood is sometimes sold by the **face cord** or **short cord**, a measure 4 feet high by 8 feet long but only as deep as the length of a single log—whether that is 1 foot, 2 feet, or somewhere in between. Buying wood by the face cord is acceptable as long as you calculate the fraction of a full cord that you are receiving and make sure that the face cord is priced accordingly.

Other measures of firewood are less precise. A truckload depends on the size of the truck. A run usually means a truckload. A rick of firewood is a stack of firewood, and it can vary considerably in volume. Thus, the best approach to buying wood is to ask for a price quote in terms of cords or cubic feet.

Find out ahead of time exactly what services you are paying for as well. Is your wood just going to be cut to length, or will it also be split? Will you have to pick up the wood from the supplier, or will it be delivered to your home? If the supplier delivers your wood, will it be dumped in a heap in your driveway, or will it be stacked neatly in your woodshed? The more services provided, the higher the price you should expect to pay. It's a good idea to be clear about price and services expected ahead of time, rather than after your supplier is driving away with your money.

Wood Quality

The single most important rule to remember about cordwood is that dry wood is superior to green wood. Green (freshly cut) wood has high moisture content, sometimes as high as 100 percent. When you try to burn green wood in a heating appliance, much of the heat generated by the fire is wasted in boiling off the moisture in the wood rather than heating your home. Burning green wood can also cause the buildup of creosote, a potentially serious problem (which I'll explain later in this chapter). Green wood has only one point in its favor: it is cheaper than dry wood. Thus, if you want to save money, the best strategy is to buy green wood in late winter or early spring and allow it to season (dry) over the summer. Some signs that wood has seasoned are: splits in the ends of the logs, bark starting to fall off, and the logs feeling lighter in weight than they were before seasoning.

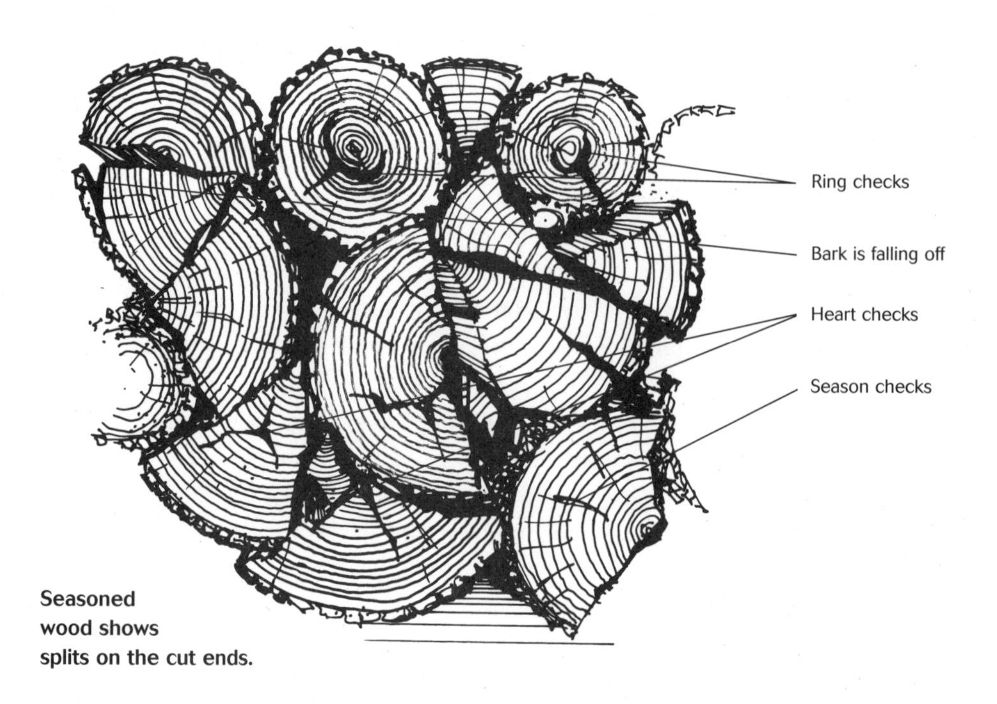

Seasoned wood shows splits on the cut ends.

Properly seasoned firewood will generally have a moisture content of 15 to 25 percent.

Because of its greater density and weight, a cord of hardwood will produce far more heat (as much as twice as many Btu) than a cord of softwood. A heating appliance full of hardwood logs will also burn longer than an appliance full of softwood logs of the same size, reducing the frequency of refueling. Nevertheless, there is nothing wrong with burning softwoods, except that they tend to produce more creosote. In some areas, softwoods are the only type of wood available. But even in areas where hardwoods are plentiful, softwoods are useful as kindling wood to help start a fire or revive one that has almost gone out. You can choose to burn softwoods on weekends or at other times when you will be at home and can tend your stove or furnace more frequently. The quick, hot fire that softwoods produce can be perfect during late fall or early spring when you just want to take the chill out of the air in your house in the morning or evening. Flexible use of a variety of woods of differing quality can help you make the best use of both your wood supply and your heating appliance.

Splitting Logs

Although hardwoods are favored for burning, they are generally more difficult to work with than softwoods, and this is particularly true when it comes to splitting firewood chunks. Splitting a gnarly hardwood like elm by hand can be well nigh impossible, while a smooth-grained softwood like poplar normally splits with ease.

Why bother to split firewood at all? The first reason is that splitting produces pieces of wood that will fit into the firebox of your heating appliance. The second reason is that splitting assists the wood-drying process. The more surface area that is exposed to sun and air, the faster the wood will dry. Another consideration is that some types of wood, especially from the birch family, won't dry properly unless they are split.

Hand-Splitting

Splitting wood by hand is a relatively quiet, clean, almost meditative procedure that is also good exercise. It's an especially effective strategy when two people are involved; one person sets up and steadies the log while the other does the splitting. If both individuals are equally skilled, they can switch jobs as desired. Hand-splitting wood is hard work, and you do need to be careful not to injure yourself (or your helper). When splitting wood, always wear eye protection.

Basic hand-splitting tools haven't changed much since the 19th century. You will need a solid chopping block that is about twenty inches tall to use as a base on which to set a log while you split it. You will also need a 6- or 7-pound splitting maul with a straight-grained hardwood handle. Two or three steel single bits (or splitting wedges) are necessary to help split logs that resist the maul. The extra wedges are useful for freeing the wedge(s) that inevitably becomes stuck in a log. You will also want a small kindling axe with a 1¾-pound head and an 18-inch handle.

The right hand tools for working with wood make a hard job a little easier—and safer.

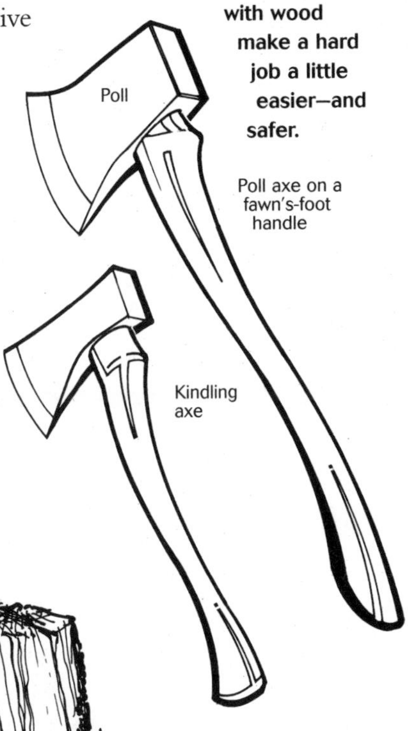

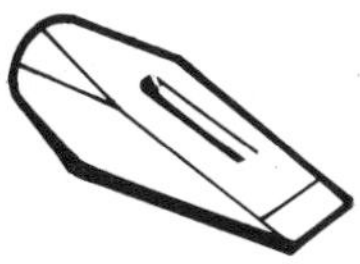

Poll

Poll axe on a fawn's-foot handle

Kindling axe

Steel wedge

6-lb. splitting maul

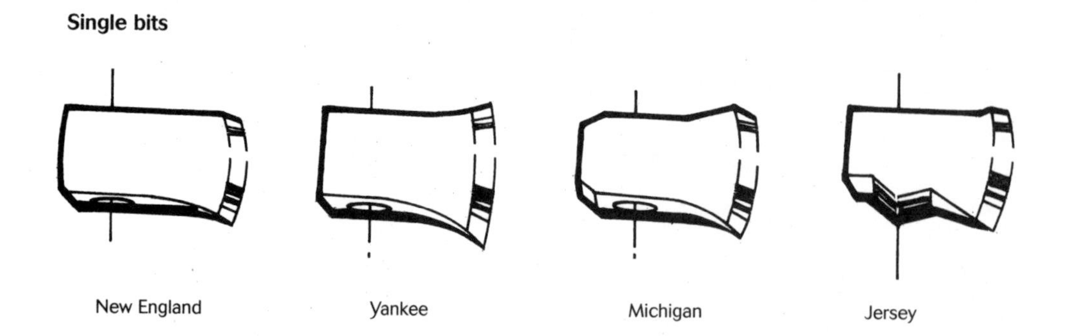

Using a Power Splitter

Splitting wood with a hydraulic splitting machine is a particularly useful strategy if you have a lot of wood to split, or if the wood is a species that is especially hard to split by hand (like green elm). It never ceases to amaze me how quickly those miserable crooked chunks of wet wood pop apart on a power splitter. The drawbacks of using a gasoline-powered splitter are the noise, greasy mess, and higher cost compared to doing the job by hand. Plus, using a fossil-fueled device to process a renewable fuel is a bit of a contradiction.

Because a power splitter is an expensive machine, it's probably better to borrow or rent one unless you are planning on splitting *a lot* of firewood over many years. Operating a power splitter is easier with a two-person crew. A four-member crew is better still. Operating this type of machine requires careful attention to safety, so be sure to acquaint yourself with the appropriate safety procedures before you begin. Eye protection, snug-fitting gloves, and steel-toed boots are a must.

Drying Wood

Wood dries when air passes over its surface, absorbs moisture, and carries the moisture away. As the moisture evaporates from the wood's surface, more moisture is drawn from the interior of the wood's cells by capillary action. When all of the "free" water has evaporated from inside the cells, the wood is at approximately 30 percent moisture. At this point, checks and splits become visible at the ends of the logs. Additional moisture can also evaporate from the cell walls of the wood cells. Split wood dries faster—nearly twice as fast—as whole logs, and smaller pieces dry faster than large pieces.

The Woodpile

In order to assist the drying process, you need to build a woodpile. Now that may sound simple, but it can be trickier than you think. Building a good woodpile is an art. It's not unusual for a woodpile constructed by a novice to topple over in the middle of the wood-burning season and especially in the spring. To avoid woodpile collapse and optimize the drying process, follow these guidelines.

- Choose a dry location away from trees and brush that might shade the pile from the sun.
- Avoid stacking the pile against the side of a building or in any other spot that would restrict good air circulation.
- Construct a foundation that will raise the wood several inches off the ground; used wood pallets work well for this purpose.
- Remove any high grass from around the woodpile.
- Stack your cut-and-split firewood with plenty of air spaces between the pieces to hasten the drying process; a crisscross pattern is best. If possible, position the pile so the prevailing winds flow through the long side.

Be sure to cover the top of the pile with a tarp, plastic, or roofing material. Old corrugated-metal roofing is my favorite because you can remove a lot of wood before you have to rearrange the roofing. Never wrap the entire pile in plastic. This blocks airflow and impedes evaporation. Your wood is ready to burn at 25 percent moisture content.

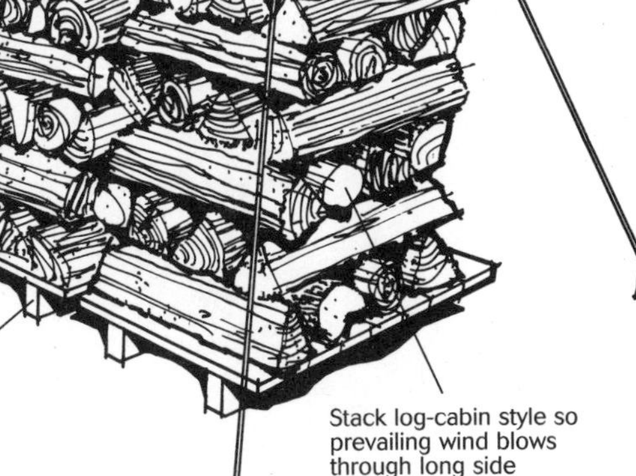

Building a stable and effective woodpile is a real art.

Handling Wood

Sooner or later, every woodburner begins to develop systems to make life a little easier. The key is to minimize the number of times you move your wood and the distances involved with those moves. Here are a few hints that will save you time and labor.

- Have your wood delivered as close as possible to the place where it will be split and stacked.
- Stack your wood as close as possible to the door through which you carry wood into your house. If you have an attached woodshed, better still.
- If your woodpile is a distance from your house, a two-wheeled garden cart is a real blessing. The big wheels make for a well-balanced, stable vehicle that can carry a surprising amount of firewood.
- Using a heavy-duty canvas log carrier or even a small hand truck can make carrying wood inside your home to the stove or furnace much easier.

Burning Wood

In order to understand how wood-fired appliances work, it's important to understand how wood burns. There are three main stages to wood combustion. In the first stage, evaporation and vaporization remove moisture from the wood. The heat required for this (above 200 degrees Fahrenheit) does not heat the appliance or the living space.

At 500 degrees Fahrenheit, wood enters the second stage and begins to break down chemically. Volatile gases form. These gases may contain 50 to 60 percent of the wood's heat value. At approximately 1,100 degrees Fahrenheit, these gases, when mixed with the right amount of oxygen, begin to burn. This temperature and a sufficient oxygen supply must be maintained to ensure complete combustion.

In the third stage, after the release of the gases, the remaining material (charcoal) burns at temperatures of 1,300 degrees Fahrenheit or higher. A small amount of ash remains after the charcoal is completely burned.

When wood is burning slowly at a

Wood combustion basics

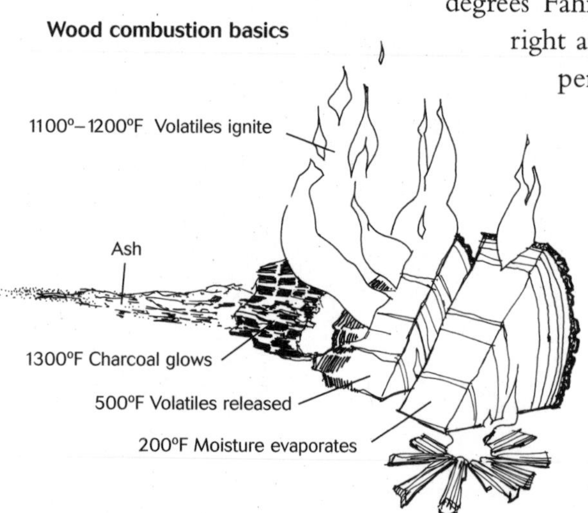

relatively low temperature, the smoke from the fire usually contains a substance known as creosote. **Creosote** includes vapors, tar, and soot, and it collects on relatively cool surfaces in the chimney flue and sometimes on the stovepipe as well. Incomplete combustion, cool surfaces, or the burning of wet or unseasoned wood (or a combination of these factors) can lead to creosote formation. Creosote not only looks and smells nasty but also can pose a serious fire hazard if it accumulates in excessive amounts. The best way to avoid creosote formation is to maintain a briskly burning fire with dry, properly seasoned firewood.

Dealing with Ashes

All wood-burning appliances produce ash. You should store the ash in a metal container with a tightly fitted lid. Place the container on a noncombustible floor or on the ground away from all combustible materials. *Never* store wood ashes in a paper bag or cardboard or wooden box because of the potential fire hazard, even if you think the ashes are cold. All it takes is one tiny live coal hidden in those ashes to start a fire.

Wood ashes are valuable, so don't throw them out with your trash. In the spring, you can add limited quantities of wood ashes to the soil of your garden (up to 20 pounds of ashes per 1,000 square feet), where they help to complete the natural cycle that is involved in burning wood. Wood ashes should not be used in your potato patch, however, because ash-sweetened soil may produce scabby potatoes. Ashes tossed in the air through the foliage of young fruit trees while the dew is still clinging to the leaves is an old-time measure that's reputed to help fight disease organisms. Ash sprinkled around young plants may deter slugs. You can also sprinkle thin layers of wood ash among other layers of organic matter as you build a compost pile.

Never put ashes in a bag!

CHAPTER 9

Woodstoves

I've been burning wood as my primary or secondary heat source for over thirty years. I've used several types of home heating appliances but mostly woodstoves. My first woodstove was a huge and ornate cast-iron creation manufactured in 1891 by Bussey & McLeod in Troy, New York. The stove was in a commercial building that I bought in the early 1970s in Bristol, Vermont. That stove consumed vast quantities of wood, but it generated a lot of heat.

From my experience with that old stove, I knew that if I wanted to heat my home without spending the rest of my life cutting firewood, I was going to have to find a more efficient heating appliance. I finally settled on an Ashley Automatic cabinet-model woodstove. In its day, the Ashley was on the cutting edge of stove technology. The stove was fired through a big door on one end, which made the fueling process easy. The large firebox even had a firebrick lining. Although some of the critical stove parts were cast-iron, the exterior was shrouded with a rectangular, brown sheet-metal cabinet that can only be described as 1960s Modern Space Heater. What the stove lacked in design aesthetics, it made up for with brute heating capacity and easy operation. The Ashley was one of the first airtight designs and sported a simple but effective bimetal thermostat that helped to control its heat output (**airtight** stoves are designed so that all air entering the stove passes through one or more controllable air inlets). This was a solid, dependable heater for off-the-grid locations like mine and for people who valued performance over looks.

Since then, woodstove technology has come a long way. Today, the two main categories of woodstoves are those that burn cordwood and those that burn wood pellets (see chapter 16 for more on pellet stoves).

Catalytic Stoves

Catalytic woodstoves contain a special smoke-combustion element that assists in high burn efficiency and low emissions. The special element is composed of a ceramic, honeycomb-like combustor that is typically located behind a baffle, out of direct contact with the flames in the firebox. The combustor is coated with a chemical catalyst (platinum or palladium) that encourages the ignition of gases at lower temperatures (350 to 600 degrees Fahrenheit). The firebox on a catalytic stove is usually larger than the firebox on a similar-size noncatalytic stove. When the combustor is operating properly, the wood burns with lower emissions. However, after a few years, the catalytic combustor wears out and must be replaced. A catalytic stove with a damaged or nonfunctioning combustor produces far more pollutants than a noncatalytic stove.

NonCatalytic Stoves

The **noncatalytic woodstove** (also known as a clean-burning or recirculating woodstove) is a far more common design. It typically includes a firebox, air controls, and baffles to recirculate smoke for more efficient combustion. If properly operated, a clean-burning stove will achieve high efficiencies and low emissions (but not as low as a properly operated, new catalytic stove). The primary combustion air in a noncatalytic stove is preheated to assist more complete burning. (The more complete the combustion, the lower the emissions.) These stoves typically have a series of baffles that keep the unburned smoke in the primary combustion zone as long as possible before sending the smoke to a secondary combustion area where reignition is facilitated by the introduction of additional preheated air.

The fireboxes on clean-burning stoves are typically smaller than on similar-size catalytic stoves to encourage fires to burn hotter. The fireboxes are often insulated or contain firebricks to help maintain higher combustion temperatures. The air inlets on these stoves often have stops that prevent them from being closed completely, which also encourages hotter fires. The disadvantage of all this is that clean-burning stoves generally do not hold a fire as long as their catalytic cousins. Nevertheless, noncatalytic stoves now account for about 80 percent of the woodstove market, according to some observers.

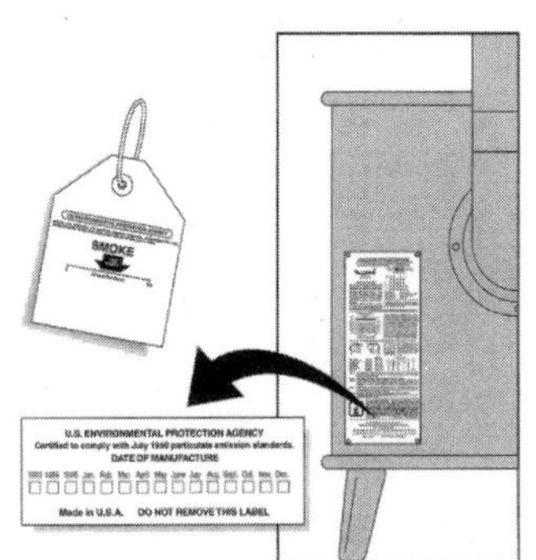

A U.S. Environmental Protection Agency (EPA) stove safety label.

Courtesy of the EPA

All woodstoves sold after July 1, 1992, are required to be certified by the U.S. Environmental Protection Agency (EPA). This certification ensures that woodstoves emit less than 7.5 grams of smoke particulates per hour (compared to about 42 grams per hour produced by stoves made during the 1970s and 1980s). All certified woodstoves now carry a permanent EPA label. The label includes the date of manufacture, location where the stove was tested for safety, and installation instructions. A second, removable showroom label provides information on the stove's emissions, efficiency, and heat output ranges.

The new woodstoves produce almost no visible smoke and deliver efficiencies in the range of 70 percent, a substantial improvement over the 40 to 50 percent ratings of earlier models.

Choosing the Correct Stove

Choosing the best woodstove can be a little like trying to choose a mate. It's something that you will have to live with on a daily basis for a long time. What's more, there is a bewildering array of possible sizes and styles, and you will be very unhappy if you make a bad choice. Also like choosing a mate, you will know whether you have the right woodstove after you spending a long, cold winter with it. Fortunately, buying a stove is usually an easier process than choosing a mate, because the factors that influence your choice are easier to quantify.

Sizing Your Stove

Technically, woodstoves are space heaters. This means that, unlike a central heating system, woodstoves have a limited ability to move hot air over long distances or through walls without assistance. Woodstoves come in a range of sizes from small to very large. Probably the most important factor in choosing a woodstove is determining what size stove meets your home's heating requirements. This is not simply a matter of noting the manufacturer's claimed heating capacity, because there are other factors involved. In addition to your home's interior layout, you need to consider its level of weatherization. The severity of the winters in your location is a relevant factor, too.

A properly sized woodstove will save you money and provide the best performance. But when it comes to woodstoves, bigger is definitely not always better. Choosing a stove that is too big is worse than selecting one that is too small, because an oversize stove is almost never allowed to burn at its most efficient operating temperature, which leads to excessive creosote formation and air pollution. An oversize stove will also make its immediate environment uncomfortably hot. At the other extreme, with a stove that is too small, you may be tempted to fire it beyond its capacity, possibly damaging the stove.

There are complicated formulas that address these issues, but you can find a basic room-heating calculator at HearthNet (www.hearth.com/calc/room calc.html). Keep in mind that while calculators are helpful, there is simply no substitute for talking with an expert.

Where to Buy

One of the best places to find a woodstove expert is at a stove specialty shop. While it is possible to buy woodstoves at hardware stores, building-supply stores, and discount chains, your best chance of getting reliable information and advice is from a locally owned stove specialty shop. There are hundreds of these operations (generally mom-and-pop stores) scattered all across the United States. Most of these retailers can provide informed advice on the stoves they carry, as well as offer delivery, installation, replacement parts, and, possibly, chimney-cleaning services. While the price you pay may be a few dollars more than you would at a "big-box" store, the added assurance of dealing with people who know what they are talking about is worth it. These specialty stores won't necessarily be the flashiest place in town. As long as the staff is knowledgeable and has had years of experience, a fancy location is not necessary.

It's a good sign when the salesperson you talk to asks you lots of questions about your house, its floor plan, and your expectations. This tends to indicate their sincere desire to help you choose the right stove for your home, rather than the one with the highest profit margin. Ask about the availability of replacement parts; if you ever need a part during the heating season, you will need it quickly. A well-stocked parts department is usually another good sign. Ask the salesperson about his or her experience with burning wood. Lack of firsthand experience may indicate that you need to keep looking for a woodstove source. Finally, don't be afraid to ask for references. A reputable stove shop should have no trouble providing you with a list of happy customers. Don't overlook talking to a local chimney sweep, who will have lots of experience cleaning up the mess caused by various types of stoves and generally can give you an unbiased, third-party opinion.

In some cases, you may also be able to buy a woodstove directly from the manufacturer. While this approach should save you some money, it also generally requires that you install the stove yourself. This is where many people get into trouble, especially if they rely on a friend or relative who doesn't know as much about safe installation as claimed. Although a six-pack of beer for your handyman neighbor may sound like an inexpensive trade for having your stove installed, if your house ends up burning down, it wasn't a smart bargain. What's more, some larger woodstoves can weigh as much as 475 pounds; trying to figure out how to handle them can be a real challenge. On the positive side, the manufacturer is the best source of information about the stoves it makes. If you do buy direct, be sure you know how to install the stove, or pay someone who is competent to do it for you. Whoever does the installation, be sure that all local fire and building codes are followed.

What to Buy

In the average stove specialty shop, the abundance of choices can be intimidating for the first-time stove buyer. For beginners, it's helpful to know that there are three main categories of woodstoves, based on their construction: steel, cast iron, and soapstone.

Russ Beamish, a knowledgeable sales representative and woodstove installer at the Chimney Sweep Fireplace Shop in Shelburne, Vermont, offers this quick introduction to the three types of stoves. "The beauty of the steel stove is that there is a very rapid heat-up," Beamish says. "We suggest them for use in basements, large areas, or rooms with cathedral ceilings where you want a lot of very intense heat created rapidly. This kind of stove will probably burn for ten to twelve hours, but as soon as the fire goes out, the stove starts to cool down very rapidly, within thirty minutes.

"Soapstone is at the opposite end of the spectrum. With soapstone, you have a lag time of heat-up because you need to saturate that inch-and-a-half-thick stone with heat before it will release into the room. Soapstone is what I suggest for people who are looking for a stove for primary heat. Initially, it will take longer to heat up, but because of its construction and heat-retention qualities, the thermal mass stays hot for several hours after the fire has gone out. So you don't get those peaks and valleys—that roller-coaster effect—as with steel stoves. I like to tell people to think of it like a freight train; it takes a long time to get up to speed and a long time to slow down. Cast iron is right in the middle. It heats up quicker than soapstone and has better heat retention than steel. It's a nice middle-of-the-road approach."

Steel

Steel stoves are made from panels cut from sheet steel or boilerplate, typically ¼- to 5⁄16-inch thick. If properly designed and constructed, steel stoves offer good resistance to warpage under extreme temperatures. Steel stoves tend to have a modern, boxlike appearance, and decorations tend to be limited to add-on ornaments or designs on cast-iron doors. One of the main advantages of steel is that the pieces can be welded together, forming a sturdy, airtight stove. Welds should be straight, smooth, and continuous; spotty welds are suspect. Many higher-quality models will have a firebrick lining in the combustion chamber. A variety of optional accessories is usually available, including heat shields, special trim packages, and mobile-home kits. Three of the best-known manufacturers of steel stoves are Avalon, Lopi, and Regency.

Steel stoves tend to have a modern, boxy appearance, and they heat up quickly.

Courtesy of Travis Industries

Cast Iron

Cast iron is the most fluid of the three materials used in stove construction. Cast iron is literally poured into a mold and consequently can be formed into virtually any shape. These stoves generally offer the widest choice of shapes, styles, and ornamentation. Cast-iron stoves are usually made from plates that interlock and can be bolted together. The seams and joints are sealed with stove cement so they will be airtight. This technique has the advantage of allowing individual parts to be replaced if they ever suffer damage (cast iron is relatively brittle and can crack if handled roughly, especially during shipping). In addition to the traditional matte black finish, many of today's cast-iron stoves are available in a variety of baked enamel colors. Optional accessories that are usually available include heat shields, special trim or leg packages, and mobile-home kits. The three primary manufacturers of cast-iron stoves are Jøtul, Vermont Castings, and Waterford.

Cast-iron stoves are often highly ornate, and they generate plentiful heat in a short amount of time.

Courtesy of Jötul North America

Soapstone

Soapstone stoves are constructed with a combination of cast iron and soapstone. The first American-made soapstone stoves were manufactured in New England in the late 1700s.

A soapstone stove offers steady, even heat and a distinctive appearance.

Courtesy of HearthStone Woodstoves

Some of these stoves are still in use today, a testament to their durability. Soapstone is quarried in blocks and sliced into sheets that are used for the stove panels. The thickness of the panels varies from one manufacturer to the next but generally is between ¾ inch and 1 ½ inches. This relatively soft stone is finished by sanding and polishing. Each panel exhibits a marblelike grain that is slightly different from other panels, making each soapstone stove a unique creation. The soapstone acts as a passive heat sink that offers steady, comfortable heat long after the fire has died down. The construction of these stoves features a cast-iron frame that holds the panels together. Gaskets and stove cement make the stoves airtight. If a soapstone panel is damaged, the stove can usually be disassembled and repaired. HearthStone and Woodstock Soapstone are the two primary manufacturers of soapstone stoves.

Beyond Materials

When you're evaluating a stove, take note of the care with which various aspects of its construction are executed, because these generally reflect on the overall craftsmanship of the stove and the quality of its design. For example, although stove legs may not be the first things you think about, they're important because they must sustain the full weight of the stove. Legs should be tall enough to allow for easy cleaning under the stove and to meet local codes for distance of the stove bottom from the floor. Make sure the legs are firmly attached to the underside of the stove. Adjustable feet on the legs for uneven floors are a nice feature.

The door system is the most frequently used movable part of any stove, and quality door design and installation are critical. Regardless of the stove type, the door is normally made of cast iron. On a steel stove, in particular, the cast-iron door provides an opportunity for ornamentation of an otherwise plain design. The door seal is extremely important, and a high-temperature gasket normally ensures that the door will be airtight. Accurate design, drilling, and assembly of door hinges are vital. Open and close the door(s) slowly to see if the hinges bind. If they do, it's a sign of questionable workmanship that may be a warning about other potential problems.

The door-handle latching system is the most important element of all, since the latch is what keeps the door shut and thus prevents burning logs from falling out of the stove. The latching mechanism should be tight enough to bring the door(s) firmly against the front of the stove. In addition, the latch should have a built-in adjustment to allow for gasket wear to ensure an airtight seal over the life of the gasket. Check all aspects of the door system carefully: it should operate smoothly and easily.

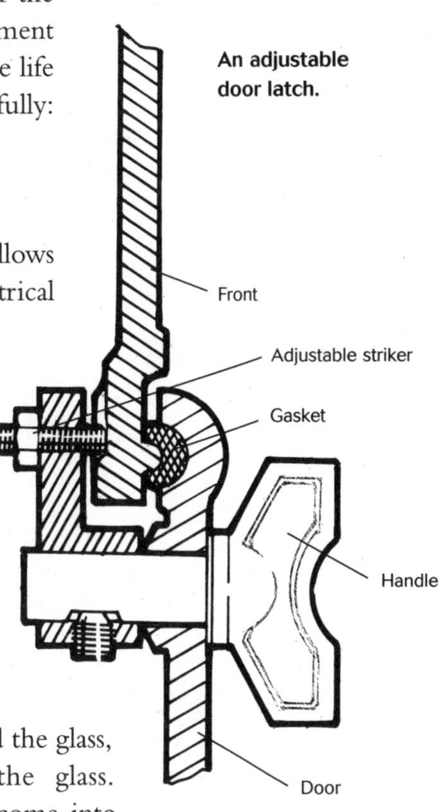

An adjustable door latch.

Other Features

Some stoves are equipped with a reversible flue, which allows for greater flexibility in installation. Blower fans and electrical thermostats are also offered as options on some models, although this strikes me as being counterintuitive for an appliance that functions so well without electricity and serves so dependably in a power outage.

Many stoves are equipped with glass windows in the door(s) that allow you to watch the fire. Being able to stare into the flames is one of the main reasons why stoves, fireplaces, and fireplace inserts are so popular; this is an important feature for most people. These doors are fitted with special heat-resistant glass and incorporate a special **air-wash** design to help keep the glass clean. This is achieved by directing air down behind the glass, creating a screen of air between the fire and the glass. Consequently, smoke or combustion particles do not come into contact with the glass, leaving it relatively clean.

Stove-top cooking is another possibility on some models of stoves. The top surface needs to be flat and to have enough space for one or more cooking pots. Some cast-iron stoves, in particular, are specifically designed to accommodate cooking, although they require special care if they are used for this purpose (consult the owner's manual for care instructions).

Many wood-fired kitchen ranges have optional **water jackets**, which are devices that will heat domestic hot water. Fitting a wood-fired home heating stove with a water jacket to produce domestic hot water during the winter months is possible, but no manufacturers currently offer this option. If you decide to pursue this strategy, be sure you find someone who knows how to retrofit a stove with a water jacket safely. An improper installation can be extremely dangerous.

Regardless of the type of woodstove you finally decide to buy, you may want to consider providing a connection from the stove's combustion chamber to outside air, especially if your home is fairly new and tightly constructed. Outside-air kits are offered as options by some stove manufacturers. This option is supposed to reduce loss of heated living-space air up your chimney and minimize the potential for combustion spillage from your stove into your home. Be aware, however, that there is considerable controversy about the use of these kits, especially with clean-burning, noncatalytic stoves. This is mainly due to the fact that the cold outside air introduced into the firebox of the stove can have a negative effect on performance. Check with your woodstove specialist before buying one.

Warranties

Be sure to read the warranty on your stove carefully, especially the fine print. There are almost as many different warranties as there are types of stoves. Some offer limited lifetime warranties, while others offer limited one- or five-year warranties on defects in materials and workmanship. Some warranties may contain a combination of these features. Some warranties are only valid for the original purchaser, while others are transferable. Almost all include a long list of exclusions. Read the exclusion list carefully. It's better to know before you buy what is or isn't covered by the warranty, rather than receiving an unpleasant surprise later on. A twenty-five-year warranty isn't much use if virtually everything that is ever likely to go wrong with your stove is excluded.

Woodstove Pros and Cons

Pros

- Wood is relatively low in cost, and when burned responsibly is environmentally neutral.
- Woodstoves can be used in new construction or renovations.
- Most woodstoves do not require electricity to operate and are a reliable source of heat during a power outage.
- Sitting in front of a woodstove is romantic.

Cons

- Good firewood is not available in all locations.
- You need a place to store firewood.
- Cutting, hauling, and using wood is labor intensive.
- Woodstoves need to be refueled several times a day, limiting the length of time you can be away from home.
- When in use, woodstoves pose a potential burn hazard to young children.
- Woodstoves tend to be somewhat dirty to operate.

Price Range

It is possible to buy a cheap woodstove from a hardware store or other retail outlet for several hundred dollars, but if you are serious about heating your home with wood, I suggest you invest in a high-quality stove. Prices for quality woodstoves vary, depending on the size and type of stove. Good steel stoves generally cost $800 to $2,000; cast-iron stoves run from $500 to $2,200; soapstone stoves are at the high end of the price range, ranging from $1,400 to $2,500. Consider a high-quality woodstove to be a lifetime investment: don't compromise on price.

Which Woodstove Is Best?

This frequently asked question is not easy to answer. If you believe all you read in brochures produced by stove manufacturers, *every* stove on the market is the best. Obviously, this is not the case. "Which woodstove is best?" is the wrong question to ask. The right question is: "Which woodstove is best for me?" From this perspective, you will soon realize that there is one stove or perhaps several that are best for your needs. All of the stoves described in this chapter will heat your home to a greater or lesser extent. You have to decide what features are the most important for you. In the end, as long as you have the correct size of stove, most of the other features are secondary; it's as much a question of aesthetics and design as anything. The stove you finally choose just needs to be the right fit for your particular tastes, usage, and budget.

CHAPTER 10

Woodstove Installation

When properly installed and operated, wood-burning appliances are no more hazardous than other home heating systems. Yet, woodstoves and fireplace inserts are the leading cause of house fires in the United States. According to the Consumer Product Safety Commission, more than 140,000 fires, 280 deaths, and 2,500 injuries related to wood-burning occur each year. These grim statistics demonstrate the importance of burning wood safely. They also indicate that there is a significant knowledge gap when it comes to heating with wood.

In this chapter, I'll provide an overview of safe woodstove installation practices. Since there can be unique circumstances involved with almost any stove installation, you should consult your owner's manual, building inspector, or woodstove professional for answers to questions about your specific situation.

Of all the renewably fueled heating appliances described in this book, woodstoves and fireplace inserts probably pose the greatest potential fire hazards, and consequently their installation and operation requires special care. Woodstoves, in particular, present multiple opportunities for accidental fires in your home. Fires can be started by heat radiated or conducted from the stove, stovepipe, or chimney to walls, ceiling, floors, and other combustible materials such as furniture or draperies. Fires can also start when sparks or coals fall out of the stove when it is opened for refueling, or when flames leak through cracks in faulty chimneys or out of the top of the chimney and onto the roof. It is even possible for sparks or coals to be blown out of a stove's air-inlet vents in the case of a chimney flow reversal (see "Effects of the Weather" on page 129). For all of these reasons, professional stove installations are strongly recommended.

Keeping the fire in your stove where it belongs and eliminating the potential for these many hazards are obviously quite important. In this chapter, we'll look at woodstove placement, clearance requirements, hearthpads, chimneys, stovepipe, and the other issues you need to be aware of to ensure safe installation and operation of your woodstove. As long as you follow the guidelines in this and following chapters, your woodstove should be a source of warmth, comfort, and security for many years.

Stove Placement

One of the first things you have to decide is where to put your woodstove. Consider several factors when making this decision. First, determine what role your stove will play in your overall home heating plan. Your woodstove can be your primary heat source, secondary heat source, or sole heat source. As a primary heat source, your woodstove will provide 50 to 75 percent of your home's heat. A backup source will provide additional heat during the coldest weather or when you will be away from home for an extended period of time. When your stove is a secondary heat source, you rely on oil, gas, electric, or solar heat for your primary heat. The woodstove is a supplement, or perhaps it heats a section of the house that your primary system cannot reach. Using your woodstove as the sole source of heat is self-explanatory, but in this case, its location is especially important.

Installing your stove in the basement is an excellent strategy for heating the basement and, to a lesser extent, the levels above. The use of floor grates will help to facilitate the flow of warm air from the basement to the living space.

A stairway also provides a good route for heat to rise. If you place your stove in the basement, plan to insulate the basement walls to reduce heat loss and moisture buildup (unless you want to use the mass of the concrete floor and walls as a heat sink, in which case they need to be insulated on the outside to keep the heat from escaping). I know of a number of people who

A Deadly Mistake

Warning! Never try to use a woodstove as a central heater by installing it under a hood connected to your hot-air furnace ductwork. This is a serious violation of building codes and extremely dangerous in the event of a chimney flow reversal. In this scenario, your home could quickly fill with combustion fumes, possibly resulting in the asphyxiation of you or your family, especially if you or they are asleep.

Fans and Air Movement

The design of many homes makes heating with a woodstove a real challenge, especially older homes that are divided into many separate rooms with little natural air circulation. This design problem can be overcome somewhat with the use of fans. A strategically placed fan can assist existing air-circulation patterns or create new ones. In order to be effective, these fans should be small, quiet, and inexpensive to operate.

One of the simplest choices is a small fan similar to the ones used to cool the electronic components of a desktop computer. They're fairly quiet, are designed for extended service, don't draw much electricity, and are easy to hang at the top of an open doorway. Despite their small size, these fans can move a significant amount of air.

Another approach is to install a wall fan to move air between two adjacent rooms. These fans come in a variety of sizes for in-wall mounting. Try to find the quietest model. Note that to use one of these fans, you will have to cut a permanent hole through the wall.

A third approach, the traditional ceiling or paddle fan, can be effective in recirculating hot air that collects in high-ceilinged rooms down to floor level. Because of their design, these fans cannot be used in rooms with ceilings less than 8 feet high.

are heating their homes successfully with a woodstove in the basement. However, it is not always easy for a basement woodstove to heat an entire house, especially a large one with many rooms (a wood-fired furnace or boiler may be a better alternative; refer to chapter 12).

Installing your stove in a central location on the first floor offers the best potential for heating all of the living spaces of your home, especially if your house has an open design. A stove located in a corner of your home or against an exterior wall will probably heat the immediate vicinity well but may leave more distant rooms cooler. This can be an advantage if you prefer, for example, cooler bedrooms. The main disadvantage of the first-floor strategy, however, is that the stove will not heat the basement, and there is potential danger that your water pipes may freeze unless you provide some other source of heat to keep them warm.

Clearance Requirements

For many years, investigations by a wide variety of government, insurance, and fire-prevention organizations have concluded that improper installation is the chief cause of stove-related house fires nationwide. The National Fire Protection Association has developed standards (continuously revised since 1906) for clearances from walls and ceilings that are used as the basis for many local building codes. All combustible materials such as woodwork, unprotected walls, furniture, and firewood should be kept at least 36 inches from a woodstove. The stovepipe should not be closer than 18 inches to an unprotected wall or ceiling.

Some woodstoves may have clearance requirements that differ from these standards; be sure to check the installation instructions and follow them carefully. In some locations, different building-code standards may apply, so be sure to inquire before you proceed with your installation. Stoves manufactured after October 1983 must have a label providing safety-related information. The Consumer Product Safety Commission requires this label; it states information about the placement and use of the stove.

Clearance distances are important because lumber that is frequently reheated will ignite at much lower temperatures than fresh lumber. A new wall or floor will not catch fire until it reaches temperatures between 500 and 700 degrees Fahrenheit, but if these surfaces are continually heated over a period of time the lumber will dry out and may eventually char due to the effects of excessive radiant heat (this includes wood framing behind sheetrock). The ignition temperature of a wall or ceiling in this scenario can drop to as low as 200 to 250 degrees Fahrenheit. It's entirely possible for an improperly installed woodstove to cause a serious disaster at some time in the future, even if it seems safe to operate initially.

There is a simple test that you can conduct to determine if a wall surface has sufficient clearance from your stove or stovepipe. When your stove is operating at a high temperature, place your hand on the area of the wall in question. If you can leave your hand in place comfortably, the wall is probably safe. If you can't bear the heat, more protection is needed. A **heat shield**, which is a noncombustible protector, may resolve the problem. A variety of commercial heat shields are available for most tight-clearance situations.

Depending on where you live and who your insurance carrier is, you may need to have a building inspector or other local official inspect your completed installation. In any case, it's a good idea to check with your insurance company before proceeding with a new woodstove installation.

Hearthpads

Back in the 1980s, there was a story in a local newspaper about a house fire in central Vermont. A woodstove without a heat shield had been installed directly on bare floorboards without the benefit of a hearthpad. Over time, the floorboards had gradually dried out and darkened. One day, while the occupants of the house were away, the floorboards caught fire and burned to the point where they could no longer support the weight of the stove, which plummeted into the basement. Fortunately for the homeowners, on its way down, the stove severed a water pipe and the resulting spray put out the fire before any further damage was done. The moral? Unless a woodstove is sitting on a bare poured-concrete floor, always use a **hearthpad**—a special insulated pad—under the stove. Hearthpads are available in a variety of styles, coverings, and colors for between $150 and $400.

Building Your Own

If you are handy with do-it-yourself projects, you can probably save some money by building your own hearthpad instead of buying one. You'll need a sheet of plywood at least ½-inch thick, two sheets of nonasbestos mineral board (or cement underlayment board), and an attractive noncombustible covering such as tile, slate, or brick. You'll frame the entire pad with a hardwood border. The pad should be large enough to extend at least 12 inches beyond all sides of the stove and 18 inches beyond the side with the loading door. Here's how to assemble the hearthpad.

1. Cut your plywood to the correct size and set it in place.
2. Cut the two sheets of mineral board so that they will completely cover the plywood.

A hearthpad.

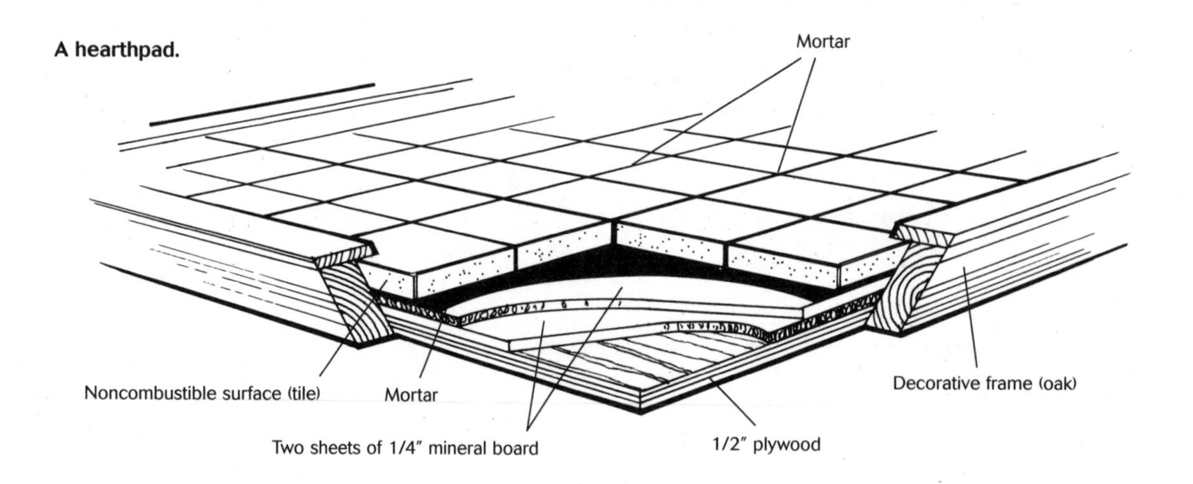

3. Cement the first piece of mineral board to the plywood, using nonflammable glue.
4. Use the nonflammable glue to cement the second piece of mineral board to the first piece.
5. Install the hardwood border around the outside edge of the plywood-and-mineral-board sandwich.
6. Use a commercial mortar mix to cement the tile, slate, or brick covering onto the mineral board. Be sure to fill the gaps between the tiles, slates, or bricks with mortar mix as well.

Let the materials harden for several days before installing the stove. If you are using a lot of brick or heavy tiles, you might want to check your floor joists to be sure they will be able to support the additional combined weight of the hearthpad and stove. If in doubt, strengthen the joists.

Chimneys

The most important part of any woodstove installation is the chimney. Chimneys that run up through the interior of your home before passing through the roof are preferred over chimneys located on the exterior of your house, because interior chimneys are less prone to creosote formation. But if your home has an existing masonry chimney, there is a good chance that you will be able to use it for your woodstove.

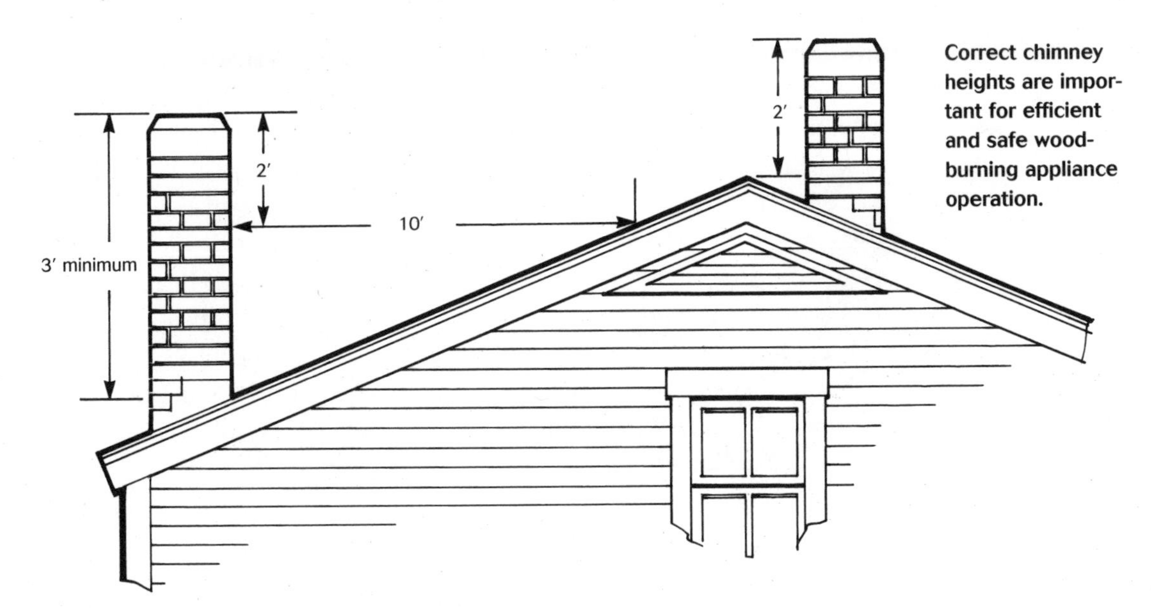

Correct chimney heights are important for efficient and safe wood-burning appliance operation.

A careful inspection—ideally by a chimney sweep or other professional—is necessary to determine if the existing chimney is safe. All chimneys must extend a minimum of 3 feet above the roof surface and 2 feet higher than any part of the building within 10 feet of the chimney. The chimney should not have any mortar or bricks missing and should have a sound clay-tile flue liner. If there is no liner, you will need to have one installed before you can use the chimney for a woodstove. A liner protects the bricks and mortar of the chimney from the corrosive properties of smoke and creosote; it also shields your home's framing lumber from the risk of fire caused by overheating or leakage of flames or sparks through defective brickwork.

Modern woodstoves need a correctly sized chimney flue in order to perform properly. An incorrectly sized liner can lead to excessive creosote buildup. The liner should be the same size as (or just slightly larger than) the stovepipe. A wide array of metal and poured-in-place chimney liners is available for this purpose. The poured-in-place liner, originally developed in Europe, has the advantage of strengthening older chimneys and insulating the new flue at the same time. Stainless steel liners are approved for wood-burning appliances, while aluminum liners (meant for gas appliances) are not. A chimney specialist can help you decide which type of liner will work best in your particular situation. In all cases, when a chimney is relined, it's important to seal the space between the existing chimney and the top of the new flue liner in order to exclude rain and wind. If you have any doubts about the proposed chimney liner, consult your local building inspector or fire department to see whether that liner is a safe choice.

Prefabricated Chimneys

If you don't have a usable chimney, the other popular alternative is the manufactured or prefabricated chimney, sometimes called Metalbestos (which is actually a brand name). Factory-built metal chimneys specifically designed for use with wood-burning appliances are often referred to as prefabricated, Class A insulated, or all-fuel chimneys. For many years, there were two main designs of prefabricated chimneys: double-wall insulated and triple-wall air-cooled. After the early 1980s, however, the triple-wall design began to fall out of favor due to problems with performance. Today the vast majority of prefab chimneys are of double-wall design. All of these chimneys have a stainless steel interior, but some offer a choice of stainless steel or galvanized outer-wall material. Although they are more expensive, the all-stainless-steel models will last longer.

If you plan to use a prefabricated metal chimney, be sure it is listed by Underwriters Laboratories. In addition, you should only use the type of metal chimney specified by the manufacturer of your stove. Other types may not be appropriate for your particular stove. Follow the manufacturer's installation instructions exactly. Most prefabricated metal chimneys must have a minimum of 2 inches clearance to all combustible materials as well as any type of insulation. A wide range of accessories is available, allowing considerable flexibility in how metal chimneys are installed. The use of a chimney cap is strongly recommended to keep out rain and birds. Prefabricated metal chimneys are not cheap, but they typically are less expensive than a new masonry chimney and are much quicker to install. In most situations, a competent installer can easily finish the job in a day.

The Fireplace Option

If your home has an unused (or little-used) fireplace along with an existing chimney, it makes sense to convert the fireplace into a real home heating system, rather than leaving it as a wall decoration. I'll discuss fireplaces in greater detail in chapter 13, but for now, I'll explain how to use your fireplace as a convenient location for a woodstove installation.

There are several ways to connect your woodstove to a fireplace. Generally, if space permits, it is best to install a woodstove on the hearth in front of a fireplace rather than inside the fireplace cavity. A woodstove heats better when all of its surfaces come in contact with room air currents; much of the stove's heating potential is significantly reduced when it is tucked back in a fireplace.

One method of making the connection between the stove and fireplace flue is to replace the fireplace damper with a special metal plate. The plate is sealed to make it airtight and a hole is cut in the middle of the plate for the stovepipe. The pipe (usually a 5- or 6-foot length of flexible stainless steel) runs through the metal plate and into the first or second flue tile of the masonry chimney. Since 1984, national fire codes and standards (as well as many manufacturers' installation instructions) have specified this type of stove-to-fireplace connection. Fireplace adapter kits, available from many stove retailers, make

A safe stove-to-fireplace installation includes a flexible stainless steel stovepipe that extends well into the flue tile.

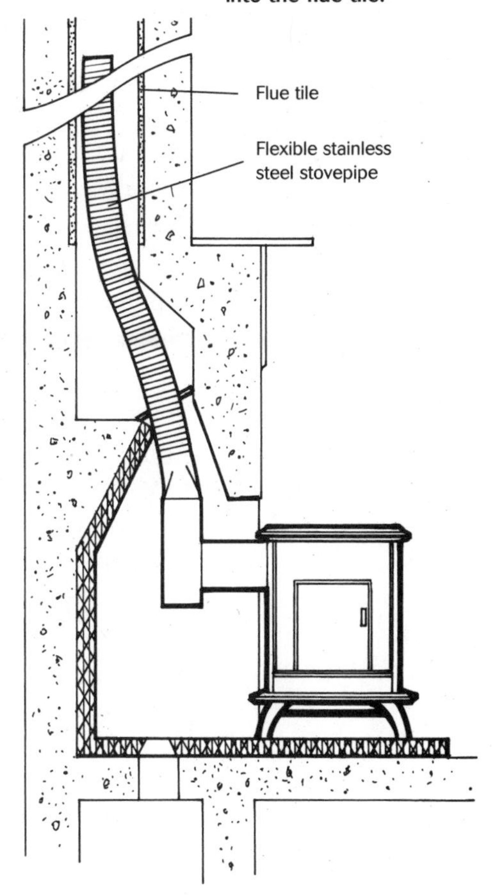

this often-dirty job much easier. Specialty woodstove retailers will perform this crucial operation for you when they install your stove.

Another installation option is to bypass the fireplace opening completely by installing a **thimble**, which is a device through which a stovepipe passes into a chimney. In this case, the stovepipe exits the stove vertically and then angles into the thimble, which is located above the mantel. Improper thimble installations are a frequent cause of house fires, so this part of the project is best left to an experienced professional. Also, if you install a thimble, you must seal the damper in your fireplace. Any part of the mantel within 18 inches of the stovepipe should be protected with a heat shield.

Whatever method you use to connect your woodstove to the fireplace, you need to be certain that the fireplace flue meets the same standards you would look for in a regular chimney. If the flue is unlined, it needs a liner before you can use the woodstove safely. If the flue is too large, you should consider installing a stainless steel liner or a poured-in-place liner similar to the ones used in chimneys without fireplaces.

Stovepipe

Your woodstove is connected to the chimney by stovepipe. This pipe is the critical connection between the two and is often the weakest link in the entire system. Many house fires are caused by improper stovepipe installation. Stovepipe should *never* pass through walls, ceilings, floors, or windows. Stovepipe may be made from sheet steel, stainless steel, or galvanized steel. The thickness of stovepipe is expressed by the term "gauge," and the lower the gauge, the thicker the metal. Thicker pipe lasts longer and resists chemical corrosion. Twenty-four-gauge pipe is generally recommended. Enameled stovepipe in colors to match enameled stoves was popular in the 1980s and early 1990s but has become less common because it is so expensive and hard to work with. Most stove installations use standard sheet steel pipe.

Keep the stovepipe run as short as possible. Eight to ten feet is the maximum safe total length. Less is better. In the 19th century, it was not unusual for stovepipes to run the full length of a room. Some public buildings such as town halls, churches, and schools had remarkably long stovepipe runs. But most stoves of that era were very inefficient and typically lost most of their heat up the stovepipe and the chimney. This allowed the stovepipe to act as a radiant heater and kept the temperature inside the pipe high enough to minimize creosote formation. This strategy does not work well with modern stoves. The highly efficient stoves manufactured today are designed to keep as much heat as possible from escaping up the stovepipe and chimney. A high-efficiency stove combined with an excessively long run of stovepipe is an invitation for disaster from excessive buildup of creosote in the pipe and chimney.

Stovepipe manufacturers crimp one end of a piece of pipe so it will fit inside an uncrimped end of another section of pipe. To provide more installation flexibility, some sections of pipe are crimped on both ends, while others have no crimp on either end. Also, you can use flexible or rigid elbows to make a wide variety of angled joints in a stovepipe. The installation of a **cleanout tee**—a T-shaped section of pipe with a tightly fitted cover over one opening—provides access for cleaning out the stovepipe and the chimney. Stovepipe comes in a variety of diameter sizes; 6-, 7-, and 8-inch diameters are the most common. Using a stovepipe of smaller diameter than the outlet from your stove's flue collar will reduce combustion efficiency and is not recommended.

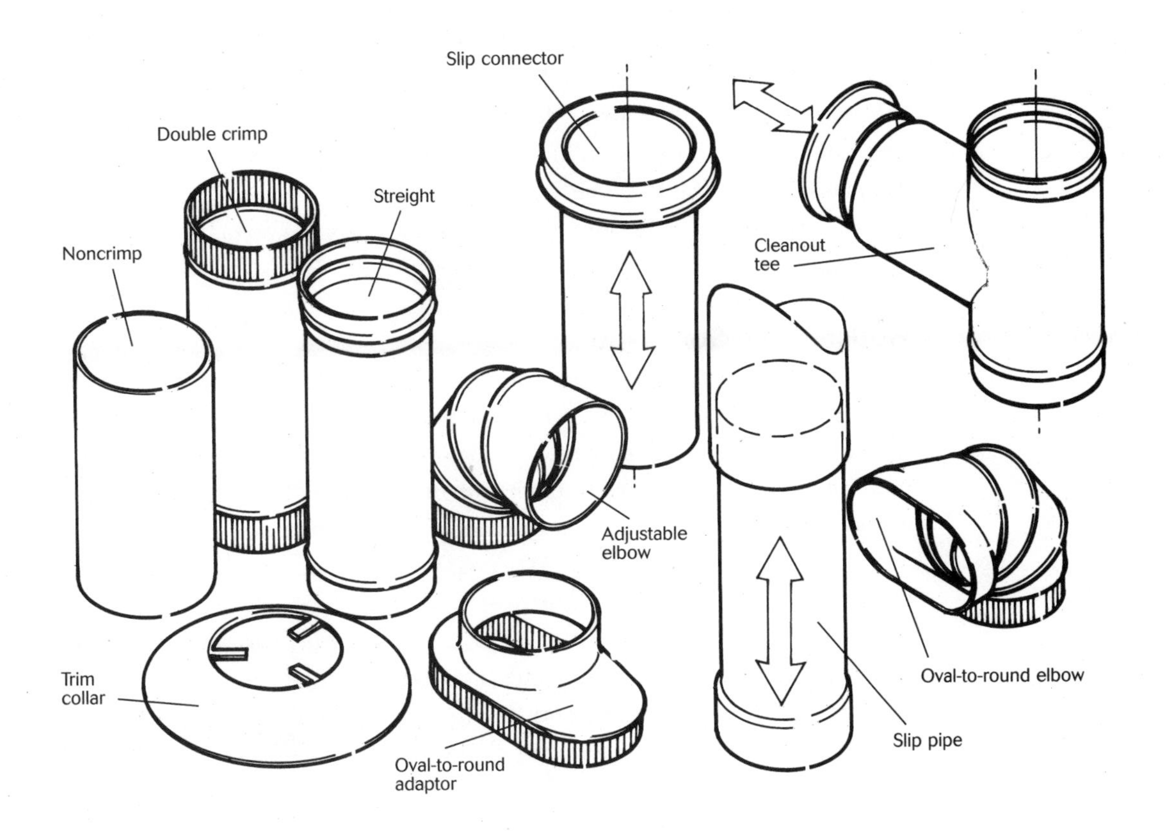

A selection of stovepipe.

How Much Do You Need?

When you are calculating the amount of stovepipe you will need, note that more is better than not enough. You can always return unused pipe to your supplier after the job is complete. The quantity of stovepipe you will need depends on the type of installation. There are three standard configurations. The first, and simplest, is an installation where the stove is directly in front of the wall thimble. For this, you will need only one section of pipe, with the crimp on the thimble side. The end that fits into the stove may or may not need a crimp, depending on your particular stove. The stovepipe should be secured with sheet metal screws to the stove's flue collar.

The second type of installation is required when the thimble is located in the wall above the stove. This calls for a vertical rise of pipe, an elbow, and then a horizontal run into the thimble. (Actually, not quite horizontal—the pipe should pitch up toward the thimble a minimum of ¼ inch per foot.) If the thimble is offset from the stove, more than one elbow may be necessary. Try to keep the number of elbows to a minimum, however. The length of the horizontal run of pipe should be kept as short as possible as well. You may encounter draft problems if the horizontal run exceeds 4 feet.

The third type of installation is where the stove is located directly below a prefabricated chimney in the ceiling. Two special pipe sections make this type of installation easier. The first, a **slip-connector**, is a telescoping section of pipe that extends from 2 to 3 feet. The slip-connector is easy to install and also to remove later for cleaning. The second special piece of pipe is the aforementioned cleanout tee. When installed below the slip-connector, this pipe provides inspection and cleaning access through its removable cover plate.

Working With Stovepipe

Anyone who has wrestled with an open section of stovepipe while trying to snap it onto another piece of pipe knows how frustrating working with this stuff can be. Interlocking stovepipe really *will* snap together, although 6-inch pipe is harder to work with than 8-inch pipe. With the right tools and a good supply of patience, completing a stovepipe installation is possible without injury or a bruised ego. Working slowly and measuring carefully are the keys to a successful stovepipe project.

Start at one end by engaging the lock. Hold it in place and then work your way down the pipe, fitting the lock together as you go. A second pair of hands is helpful. Keep in mind that the crimped ends of the pipe should point toward the stove, so that any liquid condensation will run back into

Checklist for Stovepipe Assembly

Here are the items you will need when assembling stovepipe:

- Gloves (to protect your hands)
- Sheet metal snips
- Felt-tip pen
- Tape measure
- Electric drill with ⅛-inch drill bit
- Several dozen ½-inch #10 sheet-metal screws

the stove, not out on the floor. Fasten all the pipe sections together with three sheet metal screws at each joint.

Stovepipe dampers as well as stovepipe heat exchangers that were common in the past are generally not appropriate for today's airtight stoves. Properly installed stovepipe should last for perhaps five years or more. The longevity depends on how the stove is operated and how regularly the pipe is cleaned.

If you have questions about any aspect of stove installation, your stove retailer or, possibly, your local building inspector should be able to answer them.

CHAPTER 11

Woodstove Operation

Reading the owner's manual should be mandatory for new woodstove owners. If you don't know how to operate your new stove correctly, you can actually damage it the very first time you fire it up. It's important to consult the manual and follow advice from an experienced installer. Most woodstoves require a special break-in procedure, also known as **seasoning**, during the first week or so of operation. This has nothing to do with salt and pepper, but rather with the way you initially fire the stove.

The break-in procedure for a new woodstove serves several purposes. First, the procedure helps you to become familiar with the controls and comfortable with the operation of the stove. Of equal importance, seasoning also allows the stove metal a chance to expand and contract gradually, and it introduces the metal to the stresses of repeated heating. Although the seasoning procedure is fairly standard, your particular model of woodstove may require additional precautions during the first few times you use it. Compare the directions in your owner's manual carefully to the general guidelines below.

Seasoning Your New Stove

How careful do you really have to be with your first fire, and why? "It's very important," says Russ Beamish of the Chimney Sweep Fireplace Shop in Shelburne, Vermont. "It's similar to when you buy a new car; you have to go easy for the first 500 miles. A cast-iron stove is not used to the heat, and the seasoning gives it a chance to get accustomed to high temperatures. Think of the process like a staircase; you are gradually going to increase the amount of wood that you will burn in your stove every time you build another fire, so that by your fourth fire, you will have two or three good-size pieces of wood in it. The key is to let the stove cool down completely in between firings and don't allow the stack temperature to exceed 500 degrees [Fahrenheit]."

Some woodstove professionals recommend as many as six seasoning fires in a woodstove before you fill it to capacity, while others suggest three or four. Taking the more cautious approach is generally a better idea. Your best strategy, in any case, is to follow the operating instructions that come with your stove.

Here's the procedure for making a break-in fire.

1. Open the stove's air inlets completely.
2. Set four or five crumpled pieces of newspaper in the firebox and top them with several handfuls of small kindling. Never use flammable liquids such as lighter fluid, kerosene, or gasoline (commercial fire starters specifically designed for wood fires are acceptable as long as your owner's manual does not prohibit their use).
3. Hold a burning crumpled roll of newspaper near the flue opening inside the stove for a few moments to warm the flue and improve the draft in the chimney. (If you are timid about holding burning newspaper, you can place a crumpled ball of newspaper near the flue and set it afire instead.) This step will also reduce smoke spillage into the living space during the early stage of the fire-starting process.
4. Light the newspaper and kindling in the bottom of the stove (it may be helpful to leave the door slightly ajar for a few moments as the fire starts).
5. After the fire is burning, add finger-size pieces of wood until the fire seems to be established. Then add several wrist-size sticks; after about ten minutes, add several more.

You may need to spend up to an hour gradually building the fire until it is well established and the stove is hot (but not too hot) to the touch. The firebox should be less than half full. Adjust the air inlets to a medium setting (about half open), and maintain the fire at this level for several hours. Then, without changing the air-inlet settings, let the fire go out and allow the stove to cool down completely. This ends your first seasoning fire.

You will need to repeat this procedure up to five more times with slightly larger and hotter fires each time before you finally crank up your stove to full heating capacity. These seasoning fires will cook off any packing compounds (applied by the manufacturer to prevent rust), cure the paint or finish, and generally settle the metal parts of your stove.

During your first seasoning fire, you will probably notice some potentially alarming phenomena. There will be an acrid, chemical smell and some visible smoke rising from the stove's exterior. This is an annoying but perfectly normal part of the initial break-in procedure. Open the windows and doors to allow the smoke and odor to dissipate and to prevent your smoke alarm from going off. (Obviously, it's better to season your stove before the arrival of subzero temperatures in the middle of winter.) After the first seasoning fire, your stove's finish should stop smoking, and by the third fire, the smell should have abated as well. Some people who install their own stoves actually set up the stove outside in the backyard for the first few break-in fires to avoid the smoke-and-smell-in-the-house routine. However, this may not be an option for you, especially if you are having your stove professionally installed in its permanent location. The good news about all of this is that once the break-in period is completed, it does not have to be repeated unless you have new internal stove parts installed.

For some steel stoves, for example, multiple seasoning fires may not be as critical. But, in the case of cast-iron stoves, I have heard about stoves cracking during the initial firing because proper procedure was not followed and the fire was allowed to get too hot. When you are looking at an initial investment of $1,000 to $2,500 for a stove, proper seasoning is obviously an important issue.

Starting a Fire

Once you have seasoned your stove successfully and are ready to employ its full heating capacity, what is the normal operating procedure? The main strategy during the first phase of a normal fire is to create a hot, quick-burning fire that will heat up the chimney and establish a good draft. Starting a fire for normal operation is a little different from the seasoning-fire routine. You begin the same way—with newspaper, kindling, and small pieces of wood—but then continue to add progressively larger pieces of wood until the stove is about two-thirds full. The last few pieces of wood can be full size. Warm up the flue with a newspaper torch or ball and then light the kindling. The fire should burn briskly with full flame during start-up.

With the stove door closed, you regulate the fire with the air-inlet controls to maintain the desired heat level. Don't be tempted to shut down the draft too soon; it's possible to smother the fire before it becomes fully established. Adjusting the draft takes practice. During this early stage of a fire, one of your other main goals is to establish a good bed of hot coals, which allows subsequent additions of fuel to catch fire more quickly. You know that the chimney and stove are well heated when every large piece of wood you add to the fire burns vigorously, without loss of intensity.

Starting a fire in a catalytic stove is a bit more complicated. In most catalytic stoves, there is a bypass damper that is supposed to be open when you are starting or reloading a fire. In **bypass mode**, the damper directs the combustion gases directly up the flue (bypassing the catalytic combustor) until the temperature is high enough (350 to 600 degrees Fahrenheit) to cause the combustor to "light off." A hot initial fire will help the combustor to light off more quickly. When the temperature in your stove is hot enough, change the damper setting to engage the combustor. In some stoves, the combustor will actually begin to glow at about 1,000 degrees Fahrenheit. Once lit, a combustor will continue to operate even if the combustion temperature falls to about 500 degrees Fahrenheit. A catalyst temperature monitor (which comes with some stoves or can be bought separately) will help you to use your combustor more efficiently.

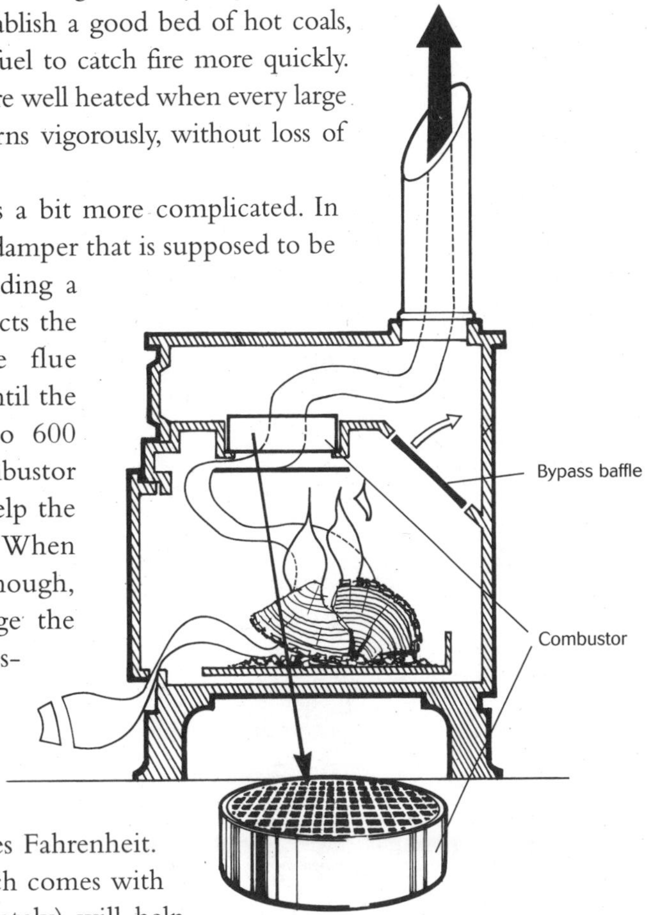

Bypass baffle

Combustor

Catalytic combustion.

Responsible Woodstove Operation

Don't burn trash in your woodstove.
Don't burn treated wood.
Don't burn driftwood.
Don't burn plastics.
Don't burn cardboard.

Do burn well-seasoned wood.
Do maintain a hot, briskly burning fire.
Do adjust the size of your fire to match weather conditions.

Maintaining a Fire

Starting a fire well is one thing; maintaining it successfully is another matter that requires practice. One of the secrets of maintaining a fire is to make sure that it is always burning with a visible flame, because a smoldering, smoky fire is inefficient and produces creosote and air pollutants. Always keep on hand a good supply of kindling and seasoned firewood of various sizes. There is nothing more frustrating than trying to revive a dying fire when all you have to work with are large pieces of wood, especially large pieces of less-than-well-seasoned wood. Having a choice of wood sizes equips you to cope with any wood-burning situations that may arise during the heating season. Controlling the heat output of your stove via the quantity and size of the wood you burn is just as important as controlling heat output through the setting of your air-inlet control. Take a flexible attitude about how you use your firewood, and you'll eventually become a master woodburner.

Another secret of maintaining a good fire, especially during midwinter, is to add more wood before the fire burns down too low. As long as a sizable bed of hot coals persists, you should not have any problems restoring the fire. Maintaining a fire requires that you pay attention to what is happening in your stove; you should check the fire every few hours if possible.

A side view of a front-to-back burner.

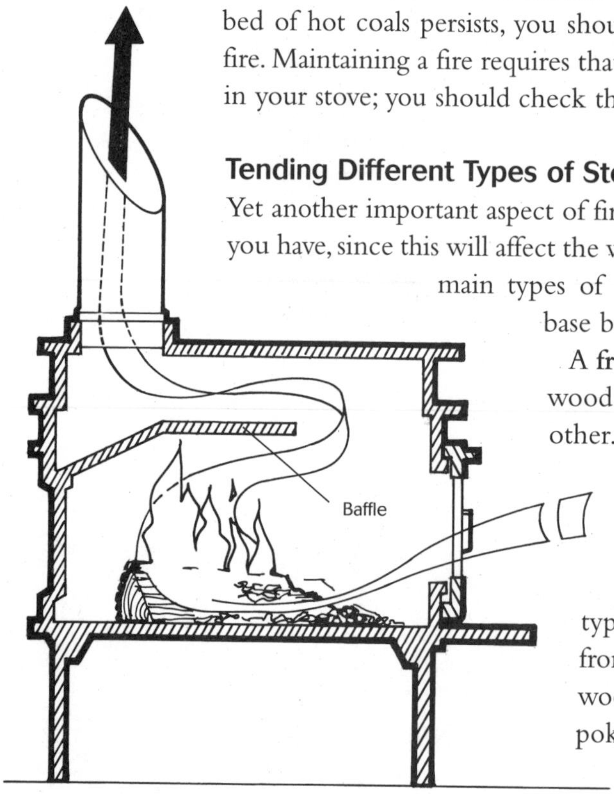

Tending Different Types of Stoves

Yet another important aspect of fire tending is to know what kind of stove you have, since this will affect the way you manage your fires. There are two main types of woodstoves: front-to-back burners and base burners.

A **front-to-back burner** is designed to burn wood like a cigar burns, from one end to the other. Your stove is a front-to-back burner if it is rectangular and has an air inlet(s) near the location where the cut ends of the logs would rest in the stove. There is a definite cycle to firing this type of stove, because the wood burns from front to rear. After your first load of wood has burned down, you need to use a poker or stove hoe to rake the hot coals toward the front of the stove (the air-inlet end) before you add the next load of wood. After adding the wood, close the door and open the air inlet fully to stimulate brisk burning. Then adjust the air to achieve the level of heat desired.

A **base burner** is generally one of the more modern types of stoves that have a bed of hot coals spread evenly on the bottom of the stove. In this type of stove, the combustion is generalized and there is no need to rake the coals from rear to front. These stoves can be either catalytic or non-catalytic. Many of these modern stoves have an air inlet just above the top edge of the window glass in the front door. This creates the air-wash feature that helps to keep the glass clean. When refueling, be careful not to place a large log on top of the front of the fire, as it might block the air-wash feature.

The Overnight Fire

Another key factor in operating your stove successfully is to know how to prepare an overnight fire—this is what separates the novices from the pros. There is a distinct strategy that will increase your chances for success. To begin with, it's imperative that you have a deep bed of hot coals in the bottom of the stove. It's also best if you have not recently added any wood to the fire. This gives you the maximum amount of room in the firebox for your overnight supply of fresh fuel. Unless your stove manual advises otherwise, pack the firebox full of large pieces of wood. Then shut the door and open the air vents fully to stimulate a brisk fire. Wait five to fifteen minutes for the fire to reach full intensity. Then, shut down the air intake to a low (but not fully closed) setting. The actual setting depends on the stove, the wood, and the weather. It may take some practice until you have a sense of where to set the air intake for a particular set of conditions. With a little luck, you'll still have a bed of coals in the morning to make the job of refreshing the fire easier.

Effects of the Weather

When you burn wood in a woodstove, you tend to become more aware of the weather, since stove performance can be affected by weather conditions. During the spring or fall, when outdoor temperatures are mild, the draft in your chimney may be reduced to the point where smoke will escape from the stove door when it is opened. To resolve this problem, close the stove door and open the air vent fully for a few minutes to allow the fire to burn briskly and warm the chimney flue.

At the opposite end of the spectrum, extremely cold weather can cause your chimney to draw too well, causing some potentially serious problems. Symptoms of unusually high draft include shorter-than-usual burn times, an excessively hot stove, and a glowing-red stovepipe. In an extreme case, the whole stove may start to glow. In this situation, you need to reduce the supply of air entering the stove. But if the stove is extremely hot, you need to reduce the air gradually, because the rapidly burning wood will continue to give off a large quantity of volatile gases that may result in **backpuffing** when these excess gases burn too quickly. These small, muffled explosions (backpuffs) in the stove can be startling or even frightening if you are not expecting them.

Operating a woodstove successfully during the spring and fall can be a challenge, because there is also a tendency for the house to overheat when you maintain a hot, efficient, nonpolluting fire. The usual response is to shut down the stove as far as possible, but this results in a less efficient fire and increased incidence of poor draft and pollution. The best strategy is to burn smaller, but hot, fires. This usually means that you will have to restart the fire in the evening (or morning), so keep plenty of extra kindling and small-size firewood on hand. And don't be afraid to open a window now and then if things get too hot. A little fresh air is nice on occasion anyway.

A chimney cap can help to stabilize drafts under windy conditions.

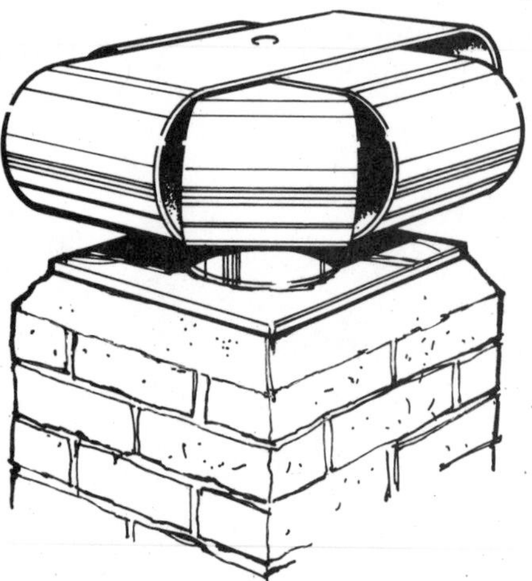

Too much wind can affect stove performance, especially if your house is located near a body of water, on a high hill, or in a valley surrounded by high hills. The wind generally causes draft problems when you are operating your stove at low heat levels. The reduced heat also reduces the upward draft in the chimney, leaving it more susceptible to downdrafts, especially if the wind is gusty. The wind blowing down the chimney—called a **chimney flow reversal**—can force smoke or even ashes and sparks out of the stove and into your living space. It can also create backpuffing.

There are several things you can do to minimize downdrafts caused by wind, the simplest of which is to burn a hotter fire. This will increase the upward draft in your chimney. A more expensive, but possibly more permanent, solution is to increase the height of your chimney, especially if it is shorter than it should be (see "Chimneys" on page 117 for guidelines on safe chimney heights). While there is no guarantee that this will work, if it does, it will solve the problem for good. A third option is to install a chimney cap. Chimney caps aren't foolproof, but if the one you install works well, you'll have no more problems with chimney flow reversal.

Chimney Cleaning

One of the most important aspects of safe stove operation is to make sure that you have a clean chimney. The Consumer Product Safety Commission estimates that about forty-five thousand chimney fires occur each year. These fires are often a result of poor maintenance and inadequate cleaning of the chimney, which leads to a buildup of creosote. Anyone who has experienced a chimney fire firsthand knows that it is an event you want to avoid. And you can, simply by cleaning your chimney. How often? The best answer is: When it needs it.

Conventional wisdom says that your chimney should be cleaned once a year. But the appropriate frequency of cleaning depends on many factors. For example, two stoves in the same house might require different cleaning schedules based on the stove type, chimney, and operating techniques. Most experienced woodburners keep a close eye on their stove installation for creosote buildup. You can check for creosote buildup using a small mirror and a flashlight, especially if you have a cleanout tee in your stovepipe. To get a general idea of stovepipe condition, try tapping on it with your fingernail or a ring. A clean pipe will produce a metallic sound, while a pipe in need of cleaning offers a muffled thud. Rap on the stovepipe sharply with your knuckles or the flat of your hand; if you hear what sounds like falling potato chips, the pipe needs to be cleaned soon.

Checking for creosote is an important task for any woodburner.

Once you have determined that your stovepipe and chimney need cleaning, you have to decide whether to do it yourself or hire someone to do it for you. There are a variety of exotic chimney-cleaning methods involving (among other things) live chickens, clotheslines, rock-filled bags, and tire chains. I don't recommend any of them.

If you have cleaned your own chimney for years, you may want to continue to clean it yourself. On the other hand, if you have no experience in chimney cleaning, it's probably best to hire a professional to do it for you, at least the first time. Chimney cleaning is one of the dirtiest jobs I know of. I cleaned my chimney for years, but eventually I decided it was worth the price to pay someone else to breathe all that miserable black dust. Besides, cleaning a chimney properly and safely requires some special tools, as well as practice. The tools—mostly brushes and flexible fiberglass rods—aren't that expensive, and you can probably recover their cost in the first year. Still, if you have a particularly tall chimney or high roof, I recommend that you hire a professional chimney sweep.

Chimney Fires

It's highly possible that, despite your best efforts to avoid it, creosote will build up in your chimney and a chimney fire will start. This happens to almost every woodburner sometime during his or her career (usually near the beginning). How will you know when a chimney fire is happening? There won't be any doubt, unless you happen to live next to a railroad. A chimney fire sounds as if a freight train is running up your chimney. It can be a truly frightening experience.

There is a classic story of a chimney fire that took place in an old unlined chimney connected to a fireplace. After the chimney fire started, a card table was used to block the airflow into the fireplace in an attempt to cut off the oxygen supply. But even before the table could be set in place, it was sucked against the brickwork of the chimney with a loud bang. Then, the table suddenly crumpled and was sucked up the chimney. This type of experience is enough to prompt most of us to invest in chimney-cleaning equipment or add our name to a local chimney sweep's regular customer list. If nothing else, it should impress you with the violent power of a chimney fire.

A chimney fire can be a terrifying experience.

A chimney fire is a severe test of your stove and its entire installation. Chimney fires resulting from creosote buildup burn at temperatures near two thousand degrees Fahrenheit—which is hot enough to damage some metal chimneys. However, if your stove is well made and properly installed, and your chimney is sturdy and in good repair, your main concern during a chimney fire will be sparks landing on the roof and setting your house on fire. Here's what to do if a fire starts in your chimney:

1. Close off all air to the stove.
2. Call the fire department.
3. Get everyone out of the house, and keep a close eye on the situation until the fire department arrives.

It's a good idea to commit this simple procedure to memory. You'll want to act without hesitation if a fire breaks out.

Never throw water on the stove or the chimney, as serious damage to both may result (you can also be injured by a sudden burst of steam). After the fire is out, have the chimney inspected for cracks or other damage before using it again. If you are lucky, the worst that will have happened is that you will have learned your lesson about the importance of regular chimney cleaning.

Maintenance

Woodstove maintenance is usually easy, and the hour or so you spend on maintenance each year is definitely worthwhile. During the heating season, you need to clean out the ashes from your woodstove on a regular basis. For most stoves, this will probably be necessary every few days. You also will need to empty out the ashes at the end of the season. Some manufacturers recommend that you leave a modest insulating layer of ashes in the bottom of the firebox.

After clearing out the ash, clean the stove's interior with a wire brush. If you have been burning hot fires, the deposits on the interior should be a light fly ash that is easy to remove. If not, you may be faced with a coating of creosote that will require concentrated work with the brush, and possibly a scraper, to remove. After cleaning, check the interior parts for obvious signs of damage or wear. The more you use your stove, the more frequently you may have to replace parts.

Check the condition of the gaskets. Look for white areas on the metal adjacent to the gaskets. These white areas may indicate isolated leaks in the gasket. You may be able to overcome a small leak by shimming a small, thin strip of gasket material underneath the existing gasket. If the gasket is dark and brittle, it should be replaced. You may be able to improve the seal of a generally leaky door gasket by tightening the door latch, if it's adjustable. If not, you will have to replace the entire gasket.

While we're on the subject of doors, if yours has a glass window that has gotten dirty, you'll want to clean it. Check your stove's manual first for guidance. There are products (available at your stove specialty store) that are specifically designed for this chore. If you use one of these products, follow the warnings on the label about wearing eye protection, gloves, and other protective gear. For best results, clean the glass when it is cool. It may be easier to clean the glass when the door has been removed from the stove or when the glass has been removed from the door. If not, clean the glass in place. In all cases, follow the directions on the product container.

Check the moving parts of your stove; they should move freely. High-temperature silica grease can be applied where needed. On some stoves, furnace cement is used as a seal on joints. Over time, the furnace cement may harden and crack out of the joints, so check the joints. Replace any missing cement with a "fingerful" of fresh furnace cement forced into the seam (wear rubber gloves). As long as you use water-soluble cement, you can wash off any excess with a damp rag.

Touch up any chips in the paint (a touch-up kit should be available from the dealer where you bought your stove). Remove rust stains from a cast-iron griddle with a wire brush and then reseason it with vegetable oil or suet (or follow the instructions in your owner's manual). Polish the handles or stove trim with metal polish, and you are all ready for the next heating season. Now you can relax—or maybe get started working up next year's firewood supply.

CHAPTER 12

Wood Furnaces and Boilers

In the 1970s, price-shocked Americans experimented with alternatives to oil for central home heating. Quite a few of them went back to burning wood. But there were other people who didn't have to go back, because they had never stopped burning wood in the first place. I remember visiting an elderly hill farmer that I knew, and the subject of heating with wood came up, as it often did in those days. "Yup, we've got a wood-fired furnace down in the basement," he replied to my question about his home heating strategy. "Fellers from the oil company been tryin' to get us to convert to oil for years, but we never did. Glad we didn't. Want to see it?" he asked, with a twinkle in his eyes that revealed considerable enthusiasm on the part of this normally taciturn Vermonter. I agreed.

Sam Daniels and one of his furnaces around 1906.

Courtesy of the Sam Daniels Company

We went into the house and down a rickety stairway into a cavernous basement that was nearly filled with enormous chunks of firewood. In the remaining space, in all its grimy glory, was an old Sam Daniels hot-air furnace. This huge woodburner—built in Hardwick, Vermont, in the early 1900s—operated entirely without electricity. When my friend swung the heavy cast-iron firebox door open, I beheld a vision from Dante's *Inferno*. The combustion chamber was huge, and it was filled with enough flaming cordwood to roast countless lost souls—or at least heat a big, drafty Vermont farmhouse. Since then, times have changed, and so has wood-fired central heating technology (although, remarkably, Sam Daniels, now located in Montpelier, Vermont, still makes wood-fired furnaces).

Wood-Fired Central Heat

Why would someone choose to install a wood-fired central heating system? One reason is that the design of some houses doesn't lend itself to efficient heating by woodstoves. Another reason is that some people who like to burn wood prefer to keep the associated mess or potential hazard for young children out of the living room and down in the basement. Perhaps the most important reason is that if you simply prefer central heating, a wood-fired system is an environmentally friendly alternative to fossil fuels. Whatever the motivation, if one or more of these categories matches your situation, you may want to consider a wood or multi-fuel furnace or boiler.

These central heating systems, normally installed in the basement or a utility room, distribute their heat via a new or existing forced hot-air or circulating hot-water system. They can heat any size home. If you're interested in wood-fired central heating, first check whether there is a ready and reasonably priced supply of firewood available in your area. Also, think realistically about how much you enjoy handling wood. A wood-fired central heating system unquestionably requires advance planning to procure fuel, plus physical labor and regular attention during the heating season. Some people love the direct personal involvement in the process, while others don't. If you fall into the latter category, consider some other heating strategy instead.

Wood-fired central heating represents a tiny niche in the overall heating market in the United States, due primarily to the generally low cost of heating oil and natural gas. But in Europe, wood-fired central heat is still quite popular in many locations. Not surprisingly, Europe has been the source of some of the most advanced central heating technologies in recent years. One of the most highly regarded manufacturers in the industry is the Danish firm Baxi/HS-Tarm.

"There have been substantial changes in the last ten to twenty years that make it easier and more convenient to use wood-fired central heat," says Lloyd Nichols, co-owner of Tarm USA, Inc. in Lyme, New Hampshire. "Part of that equation is that homes have become less energy demanding. Couple that with a more efficient, cleaner-burning heater, and the results are impressive. Now, instead of using eight to ten cords to heat a typical home, we're talking between two and five cords. We've made a very substantial reduction in the material handling required, both in terms of fuel and in terms of removal of ash. The likelihood of a potential chimney fire is virtually zero now with today's technology, because creosote in the chimney is almost nonexistent."

Nichols heats his own 2,300-square-foot home and his domestic hot water all year using only three cords of wood. "That's not something that most people think is possible with wood central heat," he says. "And you don't need to be a Paul Bunyan to heat your home with wood any more either. You can handle standard-size 18-inch wood instead of 3-foot lengths, so that anyone can easily and conveniently handle the fuel, and you are handling a lot less of it too."

Wood Gasification

Many modern wood-fired furnaces and boilers utilize a combustion process known as wood gasification to help achieve their high combustion and overall efficiencies. Simply stated, **wood gasification** involves heating the wood to drive off volatile gases, which are then burned in a secondary combustion chamber at temperatures up to 2,000 degrees Fahrenheit. These extremely hot fires create an enormous amount of usable heat and send few emissions up the stack. Except for the handling of the firewood, these heaters tend to function like their oil- or gas-fired competition and are generally controlled by a standard thermostat. When your home is warm enough, the thermostat causes the damper to close, and the fire burns at a reduced level. When more heat is called for, the damper automatically opens up again and the fire rises to a higher temperature. Many wood furnaces and boilers are installed to operate in combination with their fossil-fueled counterparts; thus, a wood furnace or boiler can become either the primary or secondary heat source for your home.

Forced Hot-Air Systems

If your home already has forced hot-air ductwork, you may want to consider installing a wood-fired furnace (as opposed to a hydronic boiler) in your home. These furnaces generally are thermostatically controlled and have an ample firebox, automatic draft control, secondary airflow, high-limit control, and optional domestic-hot-water coils. Many wood furnaces are designed to also burn coal, and some have blowers for increased combustion temperatures, as well as optional evaporation pans for humidification.

The heat provided by a wood furnace warms air in a heat exchanger. The heated air is then distributed through ductwork to the living space in your home. Because there is generally no heat-storage capacity in a forced hot-air system, the furnace needs regular refueling (on average, every eight to ten hours) in order to maintain a comfortable temperature in your home during the colder winter months. This can be a problem for people who are away from home for more than twelve to fourteen hours or who have irregular or unpredictable schedules during the winter.

Hydronic Systems

A wood-fired boiler is an obvious match for a home with an existing hydronic heat-distribution system. Most boilers feature heavy cast-iron firebox doors, automatic draft regulators, and high-limit controls. The firebox is generally surrounded by a water jacket, which transfers the heat from the fire to the water. The heated water is then circulated through the hydronic heat distribution system to your living space. In addition, the water jacket provides protection from overheating of the firebox, a design feature not available in a hot-air furnace. Most wood boilers offer domestic-hot-water coils and coal grates as options. A few offer separate hydronic heat-storage systems.

A hot-water storage tank and a wood-fired boiler are a great combination for increasing combustion efficiency and decreasing the frequency of firings.

Courtesy of Tarm USA

A hydronic, wood-fired central heating system offers many advantages. One of the most obvious is the ability to easily heat both your home and your domestic hot water from the same boiler. Another potential advantage is that a hydronic system can usually be designed to include extra heat-storage tanks that allow the boiler to be fired somewhat like a masonry heater (one hot fire a day may be all you need to heat your home). Because the fire burns quickly at a high temperature, it also burns cleanly, with virtually no emissions up the stack. The additional water tank(s) creates a heat sink where the excess heat is stored for later use. This type of storage arrangement is an excellent match with radiant-floor heating, but baseboard heaters or radiators will also work. Even with a storage-tank approach, a wood-fired boiler still needs to be refueled manually at least once a day in most situations, which can be a problem for people who are away from home for more than a day at a time during the winter.

Combination Systems

A combination wood/oil furnace offers more flexibility than a single-fuel unit.

Courtesy of Benjamin Heating Products

One way to give yourself more flexibility, while still burning wood as your primary fuel, is to install a combination central heating unit. These boilers or furnaces are designed to burn wood and either oil or gas if the wood fire dies down before you can refuel it. This type of system allows you to burn wood when convenient to do so and to rely on the backup fuel when burning wood is not convenient. Some systems are even equipped with an electric backup heating coil. Some older combination boilers or furnaces use a single combustion chamber for both fuels. Unfortunately, this creates a design compromise, and these units will never operate at peak potential efficiency for either fuel. It's also not unusual for the soot and fly ash from a wood-fired combination unit to coat the head of the oil burner, requiring frequent cleaning of the oil-burner head. This problem is eliminated when the combination boiler or furnace has two separate combustion chambers in the same unit, which is more typical of current technology. With this design, both combustion chambers are engineered to maximize the efficient burning of their respective fuel. In a multi-fuel heater, the two fuels are normally not used simultaneously. Combination heaters can be either hot-air furnaces or hot-water boilers. The main disadvantage of most combination units is that they tend to be almost twice as expensive as a single-fuel heater.

Dual Units

Dual units, also called **add-ons**, are another popular multi-fueled system. In a dual-unit system, a separate wood-fired boiler or furnace is installed as an add-on to an existing fossil-fueled central heating system. If the boiler or furnace is properly installed, a dual-unit strategy will result in extremely efficient operation for both units. This approach is an ideal way to supplement an existing fossil-fueled system with a renewable system. Dual systems offer

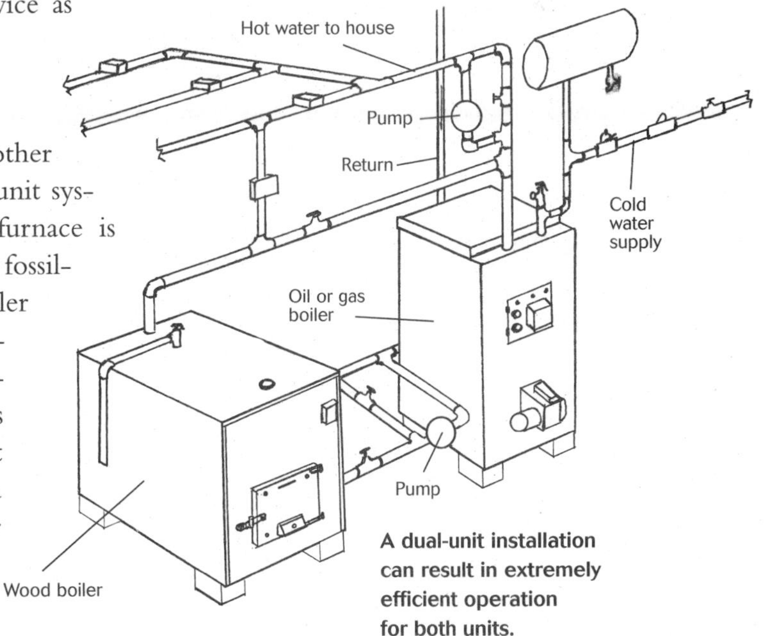

A dual-unit installation can result in extremely efficient operation for both units.

Courtesy of Benjamin Heating Products

maximum operational and fueling flexibility. If you have an older fossil-fueled boiler or furnace that is nearing the end of its career, taking the dual-unit approach allows you to burn wood while adding to the life expectancy of your old oil or gas unit. You could take this approach further by burning wood when it is convenient to do so and burning biodiesel (see "Biodiesel" on page 42 for more on burning biodiesel) the rest of the time. That would give you a choice of *two* renewably fueled central heating systems.

Outdoor Boilers

This discussion about wood-fired central heat would not be complete without mentioning outdoor boilers. See "Outdoor Boilers" on page 32 for a description and discussion of the advantages of these units. They have some disadvantages as well. Outdoor boilers tend to produce higher emissions than other wood-fired heaters, especially during the automatic restart phase after a long "off" cycle. The situation is even worse if the boiler is too large for the heating load.

According to some feedback from people who have actually used these boilers (see www.woodheat.org/technology/outboiler.htm), some of the manufacturers' claims appear to be overstated, especially those of combustion efficiencies of over 90 percent. Even if this were true, the only meaningful efficiency number is overall efficiency. The U.S. Environmental Protection Agency (EPA) sponsored a test of two outdoor boilers in 1998. The average of the tests for each unit resulted in an efficiency of around 50 percent (compared to efficiencies of about 70 percent for many EPA-certified woodstoves).

Also, claims by some manufacturers that burning a single firebox-load of wood will produce up to ninety-six hours of heat don't seem to hold up either. Around twenty-four hours appears to be more realistic. Plus, those big fireboxes on some models do consume a lot of wood. One user reported going through what he thought would be a month's supply in just one week. The longevity of some of these boilers has been disappointing; some poorly built units have been known to spring leaks after just four or five years. But most annoying of all, especially to manufacturers of other wood-fired heaters, is the claim that outdoor furnaces are somehow safer than indoor appliances. Any wood-fired heating appliance that is properly installed according to the manufacturer's instructions, that meets local fire and building codes, and that is operated responsibly is as safe as any other.

In all fairness, I've talked to a few people who love their outdoor boilers, but I suspect that emissions were probably not high on their list of priorities. If you're considering investing in an outdoor boiler, remember that it has the same disadvantages as most other wood-fired heating appliances—it's like having a barn full of cows to take care of. Of course, if you already have a barn full of cows, you are tied down anyway and an outdoor boiler might be just what you need to heat your many outbuildings. In certain situations and in locations where there are no neighbors living nearby, an outdoor boiler might be a good choice, especially if you make a concerted effort to burn properly seasoned firewood responsibly. Otherwise, if you have a modest-size home, a small wood-fired central heater or woodstove is probably a better choice. You'll burn less wood and save money in the long run.

Choosing Your System

The number of manufacturers of wood-fired central boilers and furnaces in the United States has declined substantially in recent years, due to generally low oil and gas prices. The designs, features, and quality offered by those that remain vary quite a bit. Cheaper is not necessarily better.

Unlike woodstoves, furnaces and boilers all look about the same from the outside. They generally are a rectangular box with several cast-iron doors and a few dials, gauges, or switches on the front end. Style has virtually nothing to do with selection criteria. While cabinet color may be black, blue, red, or orange, a box is still a box. Your focus should be on what's *inside* the box—how well the components are designed and how efficiently they operate. Quality of workmanship, ease of operation and maintenance are also important factors to consider.

If you are going to go to the trouble and expense of having someone drag a wood-fired furnace or boiler into your basement and connect it to your heating system, you want it to last a long, long time. If you make a bad choice, you'll be even sorrier than if you make a mistake with a woodstove. If

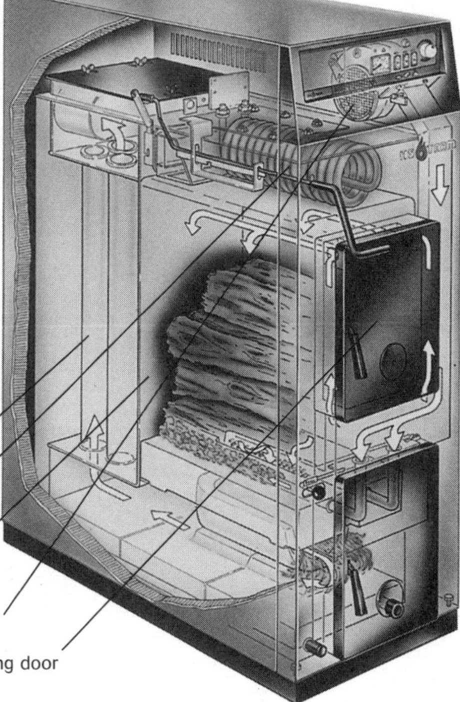

With a wood-fired boiler, it's what's inside the box that counts.

Courtesy of Tarm USA

worse comes to worst, a woodstove can be removed without too much trouble, but you will probably be stuck with your furnace or boiler for as long as you own your home.

Before you buy a wood-fired boiler or furnace, I strongly recommend that you take time to become fully educated about your choices. Talk to an experienced expert on wood central heating. If you can't find one in your area, check Web sites such as HearthNet (www.hearth.com) or woodheat.org (www.woodheat.org), where you'll find useful information, customer feedback, and industry links. Be sure to research the history of the company you plan to buy from; seek out as much unbiased, third-party feedback as possible. The longer the manufacturer has been in business, the better the chances that they produce a quality product. Some of the leading manufacturers of wood-fired central heaters are Alpha American Company, Baxi/HS-Tarm, Energy King, and Harman Stove Company.

Sizing Your System

Matching your wood-fired central heating unit to the heating needs of your home is extremely important. The amount of insulation in your home, the types of windows, and other factors will influence your choice of heater. The severity of winter weather in your area is an important consideration too. As with woodstoves, bigger is not always better when it comes to central heating units. A furnace or boiler that is too big for your needs will end up spending most of its time smoldering, creating creosote and high emissions. This is not good for your furnace or boiler, and it's not good for the environment. Your local central-heating contractor or your boiler or furnace supplier (or possibly your fuel company, if they do service and installation work) can help you determine the right-size unit for your particular needs.

Installation

While the installation of a wood-fired central heating system is definitely a job for a professional (many units weigh over 1,000 pounds and require electrical, plumbing, or sheet-metal expertise), there are some general factors to keep in mind during the early planning stages. Be sure to allow for plenty of room to work around your wood-fired furnace or boiler, and make sure you have adequate space to store your dried firewood in a handy location. The logistics of moving and handling the firewood for a central heating system is an important part of the total picture. If your basement has poor access, you almost certainly will need to improve it. Another important factor is the location of your chimney. Chimneys located on an outside wall for their entire run can be problematic for a wood-fired central heater because of the increased potential for creosote formation. Unfortunately, outside chimneys are popular with many architects because they look attractive, but most wood-fired appliance dealers and installers don't like outside chimneys. A centrally located chimney that is protected by the envelope of your home (until it passes through the roof) is the best design.

Operation

Operating a wood-fired boiler or furnace is similar to operating a woodstove. Refer to the guidelines in chapter 11 for operating woodstoves, but be sure to compare them to the directions in the owner's manual for your furnace or boiler. The same rules for preparation and storage of your firewood described in chapter 8 apply, except that it's considered best to let your wood dry for a full year before you begin to use it. This is especially true for wood-fired boilers, because the water jacket surrounding the firebox tends to remain relatively cool, and creosote will form if the wood is above 20 percent moisture content (some boiler manufacturers recommend up to two years of wood seasoning).

Another slight difference in operation is that it's not a good idea to stuff the firebox of a wood boiler or furnace full of wood in an attempt to achieve the longest possible burn time. Some furnace and boiler operation manuals specifically warn against this practice. "Filling the firebox is always the temptation, but it's best to think in terms of eight to ten hours per load," Lloyd Nichols says. "In many cases that may be just a third- or half-load of wood. If you stuff the firebox full, especially during warmer weather, you don't get an efficient or clean burn, and it's bad for the boiler or furnace because it creates acidic condensate that eats the unit from the inside."

Except during fueling, the fire door and ash door must be tightly closed while the furnace or boiler is in operation. Never burn garbage, cardboard, plywood, driftwood, kerosene, gasoline, paint thinners, used motor oil, or similar substances in a boiler or furnace. In the case of a chimney fire, shut down the thermostat and/or close the air intake vent all the way (follow the instructions in the owner's manual). Call the fire department and evacuate the house.

Maintenance

Maintaining a wood-fired furnace or boiler is more involved than caring for an oil- or gas-fired unit. In addition to feeding the fire, you need to occasionally check the grates and clean out the ash pan. Never let the ashes build up to the height of the grates, as damage to the grates is possible. Your chimney, smoke pipe, and heat exchanger will also need to be checked and cleaned at least once a year. Furnaces and boilers need servicing annually by a professional, and your wood-fired unit is no exception. Especially if your wood furnace or boiler is an add-on, have it professionally checked and cleaned every year along with the rest of the system. In any case, follow the operating and maintenance instructions that come with your unit.

Backup

Because central heating systems generally rely on electronic components such as thermostats and blowers, you must have a reliable source of electricity. If your power goes down during a winter storm, you have a problem. If you are lucky enough to have a photovoltaic electrical installation on your home with battery backup, you can wire your central heating system to operate during a power outage for a few days. But probably the best and least-expensive approach is to have a backup electrical generator. You can buy a reasonably good generator for around $600. Be sure to have a licensed electrician help you set up the system with a properly installed bypass switch; many utilities require one. Another alternative would be to install a woodstove in your basement or in a central part of your living space for use during power failures (as well as for times when you want to enjoy an open fire).

Warranties

Be sure to read the warranty on your furnace or boiler carefully, especially the fine print. Almost all warranties include a long list of exclusions. Read the exclusion list thoroughly. A solid twenty-year warranty is of greater value than a very limited three- or five-year warranty. Regardless of its duration, a warranty is worthless if the manufacturer goes out of business shortly after you buy your heater.

Price Range

While it is possible to buy a relatively cheap furnace or boiler for under $2,000, I suggest that you choose a high-quality unit that will last. Prices for quality central heaters vary, depending on the size and type of unit. The price range for an average 100,000 Btu hot-air furnace is generally $2,000 to $3,000 installed (larger units are more expensive). A complete hot-air heating system runs between $6,000 and $7,000 installed. The price range of an average 100,000 Btu boiler is $3,000 to $5,000 installed (a high-end multifuel boiler will cost $7,500 to $8,500 installed). A complete hydronic heating system runs between $6,500 and $7,500 installed (and around $10,000 for a high-end multifuel system). Consider a high-quality central boiler or furnace a lifetime investment, and don't compromise on price. Remember, once it's installed, you're probably stuck with it for life.

Wood-Fired Central Heater Pros and Cons

Pros

- They can be used in new construction or renovation.
- They will heat an entire house, regardless of its interior layout or design.
- They can provide your domestic hot water.
- They generally will hold a fire overnight.
- A multifuel or add-on unit offers fuel flexibility and security.
- Wood is relatively low in cost and when burned responsibly, is environmentally neutral.

Cons

- They usually need to be refueled several times a day, unless you have a multifueled or dual system or a large heat-storage tank.
- They tend to be somewhat dirty to operate.
- They require electricity to operate electric thermostats, pumps, and blowers.
- Good firewood is not available in all locations.
- You need a convenient place to store the firewood.
- Handling and burning wood is labor intensive.
- Sitting in front of a furnace or boiler in the basement is not romantic.

CHAPTER 13

Wood Fireplaces

Almost everyone loves to sit in front of an open fire. Warm, nostalgic visions of campfires and toasting marshmallows on sticks immediately come to mind. But I think our fascination with fire has much deeper roots that extend back to times when our very distant relatives huddled around a fire every night to avoid becoming a midnight snack for a hungry predator. It's in our genes.

"A fireplace has been, and will continue to be, the focal point in almost any living space," says Ken Rajesky, the marketing manager of Hearthlink International in Randolph, Vermont. (Hearthlink is an importer of stoves, fireplaces, and other hearth products.) "A fireplace gives you a sense of continuity or link with the past, and there is an almost primal urge to be close to it," he continues. "Watching a fire soothes your nerves and gives you an opportunity to think and relax." But this strong attraction to open fires has its drawbacks, especially when it comes to heating your home.

Brief History of the Fireplace

As I've mentioned previously, home fireplaces came into use around the 12th century. Since then, there have been minor refinements in fireplace design, but overall, sitting in front of an open fireplace today is much the same experience as it was hundreds of years ago. From a comfort standpoint, it's not ideal. On a cold midwinter night, you can overheat if you sit too close to a fireplace, but you'll feel chilled if you move too far away from the fire. Also, heating more than one room effectively with an open fireplace is a challenge. Never a very effective home heater to begin with, the fireplace sank to its lowest efficiency in the 1950s and 1960s, when both fireplace and chimney design went "modern," abandoning hundreds of years of design and construction expertise for what amounted to an architectural fashion trend. The often ugly, poorly designed fireplaces from that era continue to be an aggravation for many homeowners due to their poor performance.

During the oil crisis of the 1970s, however, many people tried to use the fireplaces in their homes as heaters. They quickly found out why previous generations had abandoned fireplaces in favor of central heating systems; the standard open fireplace just isn't efficient. At best, many open fireplaces have overall efficiency ratings of 5 to 10 percent; they consume huge quantities of firewood but produce very little heat in return. Strategies to improve heating efficiency—especially tubular heating grates, blowers, and various types of hot air ducts—generated only modest improvement. However, in the last twenty years, there has been a revolution in fireplace design and technology that has transformed the fireplace into a viable heater.

Types of Fireplaces

Despite their limitations, fireplaces are incredibly popular. While some fireplace designs are hard to categorize, there are three main types: traditional open, U.S. Environmental Protection Agency (EPA) clean-burning closed, and specialty.

The traditional open wood-burning fireplace is the type found in most homes in the United States and Canada. These fireplaces can be factory-built or of standard masonry construction. They feature an open front, usually protected by a fire screen to prevent sparks from escaping and setting rugs, furniture, or curtains afire. Because of the open front, the only way of controlling the heat of the fire is to limit the amount of fuel added, making a traditional fireplace a **fuel-limited** device. Although a traditional fireplace provides a soothing ambience, it offers little else, since most of the heat produced (about 90 percent) escapes up the chimney. This low efficiency results from the open-combustion design, which causes the fire to burn too quickly while simultaneously allowing the heated room air to be sucked up the chimney flue. In some cases, this phenomenon can cause a traditional fireplace to have a *negative* heating efficiency. What's worse, traditional fireplaces can be significant polluters, spewing up to 47 grams per hour of particulate smoke emissions into the atmosphere. It's no wonder that fireplaces are sometimes referred to as "wood toilets" in the heating-appliance industry. It's also not surprising that many western states, including (but not limited to)

Washington, Colorado, and California, restrict or ban the use of the traditional fireplace on certain days due to air-quality concerns. Some states or municipalities ban the installation of traditional fireplaces in new construction altogether.

An EPA clean-burning closed fireplace represents a significant improvement over the traditional open fireplace. Designs for these fireplaces vary. One type involves a manufactured fireplace with airtight construction that functions much like a woodstove but looks like a fireplace. After these manufactured fireplaces are installed, brick or stone facing is added around them to create the look of a traditional fireplace. Some clean-burning closed fireplaces may be exempt from EPA certification due to their weight or high air/fuel ratios. These exempt units tend to burn wood at relatively high temperatures because they admit more air to the firebox; however, they don't hold a fire for very long.

Most EPA-certified fireplaces rely on a **fireplace insert** for their low emissions and high performance. The insert is placed (or inserted) in an existing, usually masonry, fireplace opening. The EPA certifies that these inserts produce less than 7.5 grams of particulates per hour. Fireplace inserts, which are categorized by the fuel they burn, can transform your traditional masonry fireplace into a serious heating device. These inserts can burn natural gas, propane, wood, wood pellets, or coal; in this book, I'll discuss only those units that use renewables (wood and wood pellets) as fuel. We'll take a closer look at wood-burning fireplace inserts later in this chapter and at pellet-burning fireplace inserts in chapter 16.

A fireplace furnace can warm your entire home through hot-air ducts.

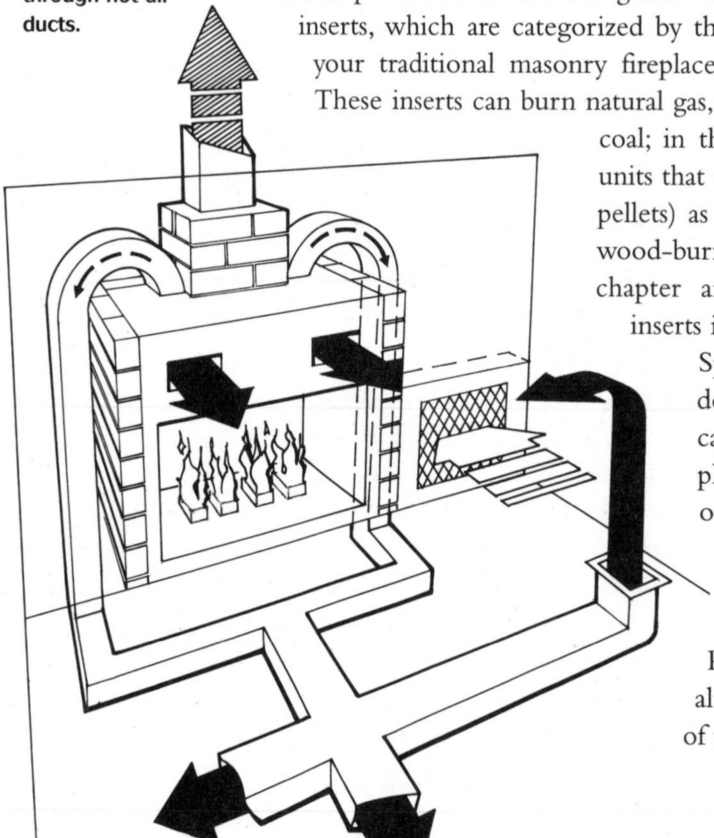

Specialty fireplaces are those that don't easily fall into the other two categories. Some of these fireplaces are EPA-certified, while others are not. One example is the Rosin fireplace. Developed in England from the 1939 scientific studies of Dr. Peter O. Rosin, this open design radically altered some of the earlier features of the famous Rumford fireplace of the late 1700s. Today's Rosin-design fireplaces, available from several manufacturers, usually consist of precast firebox parts made with industrial grade refractory material.

Another specialty fireplace is the "fireplace furnace," which is a cross between a fireplace and a central heating system. These manufactured units not only heat the room where the fireplace is located but also can heat the entire house when connected to central hot-air ductwork. Two companies that manufacture these units are RSF Energy and Country Flame Technologies (although these manufacturers may not describe the units as "fireplace furnaces").

The Tulikivi soapstone fireplace is another example of a unit that is hard to categorize. These fireplaces (from Finland) use soapstone in their construction, and although they are advertised as fireplaces, they are actually masonry heaters (more on them in chapter 14).

Location

One of the most important, but frequently overlooked, issues in home design is fireplace location. This is especially critical with masonry construction. Most architects love to put a traditional masonry open fireplace on an exterior wall because it shows off the stonework or brickwork both inside and outside of the home. While this may be nice from an aesthetic standpoint, it's the worst possible location from a performance standpoint. The massive fireplace and chimney (sometimes as much as six or seven tons of masonry) are poor thermal insulators that readily conduct heat out of your home during cold weather. If you are planning a traditional open fireplace in new construction or a renovation, locate the fireplace on an inside wall. Of course, if you are involved in a new-construction project, you should think twice before building an inefficient traditional open fireplace at all—you have so many better alternatives.

Improving the Traditional Fireplace

Before we look at the alternatives, I should mention a few things that can be done to slightly improve the performance of a traditional masonry fireplace. One strategy is to install a woodstove in (or in front of) the fireplace opening.

When a woodstove is installed this way, it is sometimes referred to as a "hearth stove" (see "The Fireplace Option" on page 119 for details). With this approach, your fireplace loses its original identity, becoming merely the convenient location to install the stove and connect the stovepipe.

One simple step you can take to reduce heat loss slightly from a traditional open fireplace is to ensure that the damper seals tightly when you close it. Old dampers are often warped, and when they are, valuable heat escapes up your chimney even when you aren't using the fireplace. If you find that the damper in your fireplace is warped or otherwise unsound, have it replaced.

An alternative way to reduce heat loss from the flue is to install a top damper on the top end of your chimney flue. A top damper is a unit—operated by a long stainless steel cable that runs up the flue—that completely seals the flue when the fire is out. A top damper also keeps rain, animals, and debris out of the flue. A top damper can *only* be used on a wood-burning fireplace flue. (Warning: Never close the damper in a fireplace until the fire is completely out.)

You can improve the efficiency of your fireplace about 10 percent by adding a set of glass doors.

Courtesy of Portland Willamette

Installing glass doors will also improve the efficiency of your masonry fireplace (by perhaps 10 percent). Although the glass in the doors reduces the amount of radiant heat flowing into the room, there is a net heating gain because the doors substantially reduce the amount of heat loss up the chimney. If you have an open fireplace, and want to use it as such, I strongly advise installing a glass-door unit. If you have a factory-built fireplace, however, check first to see whether it is approved for use with glass doors.

Another way to slightly improve heating efficiency in a traditional fireplace is to install a heat exchanger. These devices usually involve the installation of large-diameter metal pipes or tubes either in the chimney or around the firebox area. The idea is that the tubes will extract heat from the fire, and the heat will circulate through the tubes into the living space. Some of these heat exchangers use fans to move the air, while other models rely on natural convection. In either case, adding a heat exchanger only improves efficiency by 5 to 10 percent.

One final possible strategy is to install a stainless steel or decorative enameled cast-iron fireback to reflect radiant heat out into the living area. However, an easier and better approach would be to install a fireplace insert or factory-built fireplace.

With the use of an EPA-certified insert or factory-built fireplace, you can have virtually the same experience as sitting in front of a traditional fireplace, while heating your home at the same time.

Fireplace Inserts

A far better strategy for improving your existing traditional fireplace is to install an insert. A fireplace insert can transform an inefficient open fireplace into an EPA-certified heater. In most respects, a wood-burning fireplace insert is designed to function like a woodstove. Like woodstoves, fireplace inserts are made from steel or cast iron and generally feature self-cleaning glass doors that allow you to watch the flames. Those closed, tight-fitting, gasketed doors significantly improve combustion efficiency and heat retention in your living space. The closed-door system slows the rate of fuel consumption and increases the temperature for more complete combustion. Most inserts have a built-in air circulation feature as well as manual or thermostatically controlled air inlets. Some units have catalytic combustors, while others rely on clean-burning technology to achieve overall efficiency ratings similar to those of freestanding woodstoves.

A fireplace insert can transform an inefficient open fireplace into a functional home heater.

Courtesy of Osburn Manufacturing, Inc.

When you burn properly seasoned wood in an EPA-certified wood-burning fireplace insert, emissions will be almost zero. Some fireplace inserts include options such as fans and thermostatic controls, as well as a variety of aesthetic choices such as decorative tile, enameled fronts, and brass or gold trim. But best of all, you can still sit in front of a burning fire and soak up delightful radiant heat while the insert also warms the entire room and perhaps much of your home as well.

Design

Internal designs for fireplace inserts can be single wall or double wall. In a single-wall insert, the room air is drawn underneath the firebox and then rises between the back of the unit and the masonry walls of the fireplace. Now heated, the air flows over the top of the unit and back into the room through a vent. In a double-wall insert, the air is drawn into a space between an interior and exterior wall of the unit, where it is heated. The heated air flows back into the room, usually through a vent at the top front of the unit. The double-wall design is more common, safer, and also more efficient than the single-wall variety. Some inserts use a fan to circulate the heated air, while others rely on natural convection. Inserts that rely on manual controls and natural convection are obviously more useful during a power failure than those that rely on electricity.

Installation

Proper installation of a wood-burning fireplace insert is very important. Inserts cannot be installed in many factory-built fireplaces. If you're not sure what type of fireplace is in your home, ask a professional installer to inspect your chimney and fireplace to ensure that they are suitable before you proceed with an insert project. In the past, inserts were installed in the fireplace opening without any direct connection between the insert and the chimney flue. This practice led to excessive creosote buildup and an increased incidence of chimney fires. To prevent this, a direct connection between the insert and the flue tile of the chimney is now required by national fire codes and is specified in most manufacturers' installation instructions.

The same general rules that apply to chimneys and woodstoves apply to fireplace inserts as well. If your chimney is unlined or has an oversize flue, you will need to install a stainless steel liner or a poured-in-place liner (see "Chimneys" on page 117). When the chimney connection is complete, the insert is then installed in the fireplace opening and any remaining space between the sides and top of the insert is covered with a metal shield (or shroud) to seal the opening. Only approved insulating materials should be used to seal gaps between the fireplace opening and the insert shield.

Operation

Operating a wood-burning fireplace insert is about the same as the procedure for a freestanding woodstove. Like a woodstove, an insert needs to be seasoned during the first half-dozen fires before being heated to its full capacity. Never burn garbage, cardboard, plywood, driftwood, kerosene, gasoline, paint thinners, or similar substances in an insert. Follow the operating instructions in your owner's manual and the general suggestions on woodstove operation in chapter 11. Also be sure you've committed to memory the steps to take in case of a chimney fire (see "Chimney Fires" page 132).

Maintenance

As with any wood-burning appliance, you'll need to remove the ashes from your fireplace insert on a regular basis, probably every few days. Remember to store the ashes in a metal container with a tightly fitted lid. If the glass window in your insert door becomes dirty, clean it following the instructions listed under "Maintenance" in chapter 11. A fireplace insert and the chimney to which it is attached should be inspected and cleaned annually to avoid an excessive buildup of creosote. In order to do this, however, the insert generally needs to be removed from the fireplace opening. This can be a difficult operation, and you may want to assign this chore to a professional chimney sweep. Some inserts are designed with wheels to simplify installation, cleaning, and maintenance activities. Follow the manufacturer's maintenance instructions that come with your insert. Travis Industries, Quadra-Fire, Harman Stove Company, Osburn, and Vermont Castings are some of the most prominent manufacturers of wood-burning fireplace inserts.

Factory-Built Fireplaces

Another way to get better performance out of a fireplace is to start with a factory-built fireplace. Generally speaking, a **factory-built fireplace** (also called a zero-clearance fireplace) consists of a lightweight metal firebox connected to a metal chimney. Some of these units are sold as a package that includes the fireplace, metal chimney, and miscellaneous installation hardware. If the fireplace is not sold with a chimney, then the type of chimney specified in the installation instructions *must* be used.

"A well-built and well-designed manufactured fireplace offers many advantages," Ken Rajesky says. "You don't need heavy-duty structural support underneath; you can install a factory-built fireplace almost anywhere at almost any time; you can easily dress it up in any way you want, and the chimney is easily installed. When these fireplace units are completely installed, it's extremely difficult with some models to determine whether it is factory built or masonry."

A factory-built fireplace can look virtually identical to a real brick-and-mortar version.

Courtesy of Heat-N-Glo

Design

The diversity of design and features on factory-built fireplaces can be extremely confusing. Many of these units, often referred to generically as **heatilators**, are steel chambers that include hot-air ducts in their design. (Heatilator is actually the name of the manufacturer that patented the first air-circulating fireplace in 1927; they still build fireplaces today.) Some units have fireboxes lined with firebrick or other refractory material, while others do not. Some factory-built units have tight-fitting glass doors, some don't. Some use fans to circulate heated air, while others rely on natural convection. Manufactured fireplaces can be either EPA-certified or EPA-exempt, depending upon their design. The EPA-certified units tend to burn longer and use firewood more efficiently than the exempt models. But regardless of their design, most factory-built fireplaces are somewhat more energy efficient (up to 67 percent versus around 10 percent) and provide more heat than their traditional open masonry counterparts.

Installation

Many factory-built fireplaces are fairly lightweight, relatively inexpensive, and easy to install. Since they weigh only 300 to 600 pounds, most of these units do not require heavy concrete slabs or footings. Another advantage is that most factory-built fireplaces can be installed within a few inches of walls and wood framing (hence the name zero-clearance fireplace). Unlike a typical masonry fireplace, installing a factory-built unit is not beyond the ability of an experienced do-it-yourselfer. (It is critical to follow installation instructions to the letter.) Most people, however, will want to leave this task to a qualified professional. Once installed, a factory-built fireplace can be faced with a wide range of materials, including brick, tile, marble, granite, or other decorative stone.

A factory-built fireplace will wear out much sooner than a properly constructed masonry fireplace. Also, if a factory-built unit breaks and replacement parts are no longer available, you will have to replace the entire unit.

Operation and Maintenance

Operation and maintenance instructions for factory-built fireplaces vary, so be sure to follow the instructions that come with your particular unit. A few general rules do apply, however. Most factory-built fireplaces need to be seasoned just like a woodstove (see chapter 11 for the procedure). Unlike a rugged masonry fireplace, which can take a lot of abuse, overfiring a factory-built fireplace can cause severe damage to the fireplace and, in the case of a chimney fire, the metal chimney as well. The same fuel prohibitions that apply to a fireplace insert also apply to a factory-built fireplace. Serious damage to the unit may result if these prohibitions are ignored. Travis Industries, RSF Energy, Heatilator, Majestic, and Heat-N-Glo are some of the best-known manufacturers of factory-built fireplaces.

Warranties

Warranties on fireplace inserts, and especially on factory-built fireplaces, vary widely and can run anywhere from five to thirty years. Be sure to read the warranty carefully, especially the fine print. Almost all warranties include a long list of exclusions.

Where to Buy

A stove-and-fireplace specialty shop is the best place to go for information on fireplace inserts and factory-built fireplaces. Not only will the employees in a specialty shop be able to provide you with competent advice for your particular situation and needs, but also they will be able to professionally install the unit you decide to buy.

Price Range

Wood-burning fireplace inserts cost between $1,200 and $2,300. Add another $400 to $750 for installation and $800 to $1,200 if you need to reline the chimney flue. There is a wide range in prices for factory-built fireplaces. Well-built basic units start at about $900 and can run to more than $5,000 for top-of-the-line models. Add another $400 to $600 for the metal chimney and $400 to $750 for installation of the fireplace and chimney. You will also need to hire a mason to install the stone or brick facing around the fireplace unit. This can cost around $500 (or much more), depending on the materials and design of the work. Overall, expect to pay between $4,000 and $5,000 for a high-quality factory-built fireplace, completely installed. While you can pay less, many of the cheaper units are not designed to be primary home heating systems. If all you want is an occasional evening by the fire, a cheaper unit is sufficient. However, if you really want to heat your home with a fireplace, choose a reliable manufacturer and buy a well-made unit that is specifically designed for home heating.

Fireplace Pros and Cons

Pros

Traditional fireplaces can be included in new construction or renovations (new construction is generally easier).

Manufactured fireplaces can be used easily in new construction or renovations.

Many fireplaces do not require electricity to operate and, depending on the model, can be a reliable source of heat during a power outage.

Sitting in front of a wood fireplace, regardless of its design, is romantic.

Cons

Good firewood is not available in all locations.

You need a place to store the firewood.

Burning wood is labor intensive.

Traditional fireplaces cannot be used under certain atmospheric conditions in certain states.

Traditional fireplaces are heavy and require adequate footings and foundations.

Traditional fireplaces need to be refueled frequently and are generally not clean burning.

Clean-burning and EPA-certified fireplaces need to be refueled more than once a day.

Fireplaces are extremely hot and pose a potential burn hazard to young children.

Fireplaces tend to be somewhat dirty to operate.

Most fireplaces are only room or zone heaters and must be backed up by some other form of heating system.

CHAPTER 14

Masonry Heaters

Wouldn't it be nice if someone invented a simple home heating appliance that provided steady, comfortable, radiant heat and didn't depend on electricity, fossil fuel, or constant attention? Someone already has: the masonry heater. Developed between the 17th and 18th centuries in northern Europe, the masonry heater is one of the most effective wood-fired heating appliances ever devised, and its outstanding features have withstood the test of time. The masonry heater is also called a Kachelöfen or tile stove, as well as a Russian, Siberian, Finnish, or Swedish stove or fireplace. Masonry heaters are still extremely popular in many northern European countries but have not been widely accepted in the United States, due in large part to our continued overreliance on fossil fuels. But that's beginning to change.

Masonry Heaters in North America

In the late 1970s, masonry heaters were introduced in North America and since then have gained a small, enthusiastic, and growing number of devotees in both the United States and Canada. As local masons and homeowners learned about the benefits of masonry heaters, a small group of converts organized the Masonry Heater Association of North America (MHA). Made up primarily of experienced heater masons, manufacturers, dealers, and installers, this group has worked tirelessly to spread the word about masonry heaters and to develop performance standards for the industry. In 1998, the MHA established a rigorous heater-mason certification program.

Known variously as masonry heaters, masonry stoves, or masonry fireplaces, these devices are all basically the same and operate in a similar manner. "It's like having a big warm friend in your living room," says David Lyle, of Acworth, New Hampshire, author of *The Book of Masonry Stoves* (Chelsea Green, 1984). "A masonry stove is comfortable to live with, and it can be very beautiful and adapted to your own taste and style.

"Masonry stoves are not terribly well known in this country," Lyle continues, "but I think as the number of these units gradually increases, more people will grow fond of masonry stoves once they get used to their many positive features. With all the scares about fuel price swings and shortages, having a masonry stove in your home is very reassuring." Currently there are eight thousand to ten thousand masonry heaters in North American homes, according to several industry experts, so there is room for additional growth in the market.

A masonry heater with a glass door offers the visual appeal of an open fireplace and the heating advantages of a masonry heater.

Courtesy of Biofire, Inc.

How They Work

Masonry heaters are designed for quick, hot fires in their combustion chambers. The heat from these intensely hot fires (1,300 degrees Fahrenheit or higher) passes through a series of baffled chambers (heat exchangers), is absorbed by the large thermal mass of the heater, and then radiates slowly and gently into the surrounding living space for many hours after the fire has burned out. The radiated heat is absorbed by the floor, walls, ceiling, furniture, and even people, creating an extremely comfortable environment with very even temperatures.

Most masonry heaters have doors that are designed to provide optimal air supply for the fire during the burn cycle, which typically lasts about two hours. The rate of air intake is carefully controlled. Except during fueling, the heater doors remain closed, although a few models can operate with the doors open (in which case, the heater looks and functions more like a fireplace). Some doors have glass windows that offer views of the fire while the door is closed.

Masonry heaters are generally fueled once or twice a day with bundles of relatively small-diameter wood for quick combustion. Heating efficiencies can be up to 90 percent, with very low emissions. The amount of heat produced is controlled by the amount of wood burned, making masonry heaters fuel-limited devices. Except when the fire is burning, masonry heaters are silent, so they are a particularly good choice for people who are sensitive to mechanical noise. Masonry heaters work best in homes with open floor plans, but can easily heat individual rooms. Generally speaking, a masonry heater installed in the basement does not heat a house well.

Masonry Heater Design

Especially in northern Europe, masonry-heater designs are incredibly varied. These heaters range from simple whitewashed-clay models, which have served as baking ovens and home heaters, to ornate tiled masterpieces that warmed the castles and palaces of European royalty.

One of the most popular types of masonry heaters is based on early Finnish and Swedish contraflow designs and is often called a Finnish fireplace. **Contraflow** refers to the fact that hot combustion gases first flow up from the firebox to the top of the stove and then back down through several heat-exchange channels. Other designs based on eastern European traditions contain a series of vertical up-and-down smoke channels located above the firebox. Still other designs were developed in various countries in response to local needs. There have been long-running debates among European masonry experts about which design is best. Some of these controversies have been transplanted across the Atlantic. While design differences may be significant to a stove designer, the average person will find that almost any design will work effectively as a home heater.

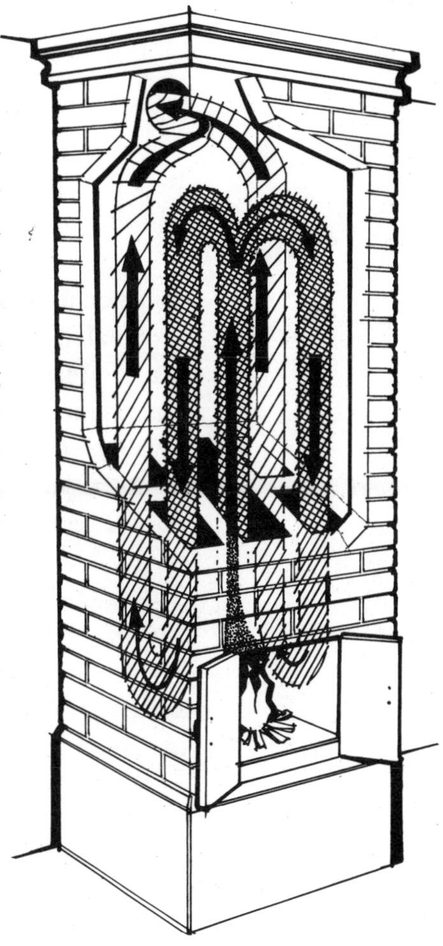
A typical contraflow stove design.

Fuel

Although there are masonry heaters that will burn other fuels, I'm going to focus on wood-fired units. Because you only need to build one or two fires a day in a masonry heater, wood consumption is generally less than for most other wood-fired appliances. An average masonry heater will consume only about two or three cords of well-seasoned firewood per winter. What's more, masonry heaters can use a remarkably wide range of wood as fuel. Hardwood, softwood, lumber scraps, small branches, or even bundles of twigs can be burned successfully in a masonry heater.

The best firewood size is pieces three to five inches in diameter, rather than the larger firewood used in a conventional woodstove. Although a few pieces up to six inches in diameter can be added once the fire is well established, it's best to use uniformly-sized, smaller-diameter wood so that it will burn down to coals all at about the same time. This means that you may have to split your wood into slightly smaller pieces of consistent size. On the other hand, you may be able to get a good deal on "limb wood" from your firewood dealer, because most other customers won't want it. Like all other wood-burning heating appliances, a masonry heater needs well-dried (about 20 percent moisture content) firewood to operate efficiently. (For guidelines on storing and drying wood, see chapter 8).

Design and Construction Considerations

Building a traditional masonry heater from scratch is a time-consuming, complicated procedure. Just the core of a masonry heater can contain five hundred or more refractory bricks, to say nothing of the many hundreds of ordinary bricks on the exterior. The internal design of the core can be extremely complex and, if it's not built correctly, can result in poor performance or even structural failure. For all of these reasons, only a professional mason who is both familiar with and experienced in construction of traditional masonry heaters should assemble these units. Check with your local housing inspector or other building officials to see if there are any local building codes that might affect your masonry heater project before you begin construction.

Sizing and Mass

The size and design of your home, as well as the climate in your region, will largely determine the size (mass) and surface area of masonry heater you need. Your heating expectations and intended usage are the other key factors. Here are some rules of thumb to consider in sizing a heater.

- The greater the mass of a masonry heater, the larger the heat storage capacity.
- The thicker the walls of the heater, the slower the response time (the speed at which it delivers heat).

- Colder climates call for more-massive heaters in order to store more heat, while milder climates call for less-massive heaters.
- In rough terms, a masonry heater should have four to eight times the surface area of a cast-iron stove in order to provide an equivalent heat output.
- The thicker the heater wall, the larger the surface area required.

Refer to *The Book of Masonry Stoves* if you're interested in formulas that help you calculate size and mass for a masonry heater.

Regardless of their size, masonry heaters are heavy. The heater core alone can weigh one to four tons, and the combined weight of the core, façade, and chimney can easily be ten tons or more. These heaters require substantial footings and foundations to properly carry the weight. Although it is possible to install a traditional masonry heater in an existing home, it's obviously a much easier project in new construction (masonry heater kits, described later in this chapter, are more compatible with renovations).

Chimneys

Most traditional masonry heaters are built in conjunction with a masonry chimney. The chimney can be an existing unit or one that is built along with the masonry heater. If you are going to use an existing chimney, have it professionally inspected first to ensure that it is suitable. The chimney should be located beside or behind the heater. Never build a masonry chimney on top of a masonry heater. In addition to the enormous structural stress on the heater involved with this design, if anything should ever go wrong with the heater, the entire chimney will have to be dismantled. If a prefabricated chimney is used, be sure to connect the chimney to the heater using an approved masonry adapter. Under certain circumstances, a masonry chimney can be combined with a prefabricated metal chimney at the top of the chimney. An experienced masonry-stove installer is the best guide for this type of question. The best location for the chimney is inside the envelope of your home, rather than on an outside wall.

Dampers

There are two main schools of thought about the use of dampers in masonry heaters. One viewpoint, represented by countries in extreme northern Europe (such as Sweden, Finland, and Russia), is that the damper can be completely closed when the fire is out. In other European countries (Germany, Austria, and Switzerland), dampers that close completely are banned. In these countries, an 80 percent damper closure is recommended. The same debate rages on in the United States as well. Nevertheless, the use of a top damper on the top of your chimney is generally recommended for masonry heaters. Once the fire has burned out *completely,* the top damper can be closed to prevent heat from escaping up the chimney. If you use a top damper, however, I strongly recommend that you also install a carbon monoxide detector as a safety measure in case the damper should ever be closed too soon.

Exterior Finish

The core of a masonry heater may be covered with a wide variety of façades or finishes, including brick, tile, soapstone, stucco, slate, marble, granite, and fieldstone. The general recommendation for thickness of the exterior façade is from four to six inches. If a heater façade is thicker than six inches, it becomes more difficult to anticipate how to respond to changing weather conditions because of the long lag time until the heater warms up or cools down. A thicker façade does offer the advantage of more heat storage mass, however.

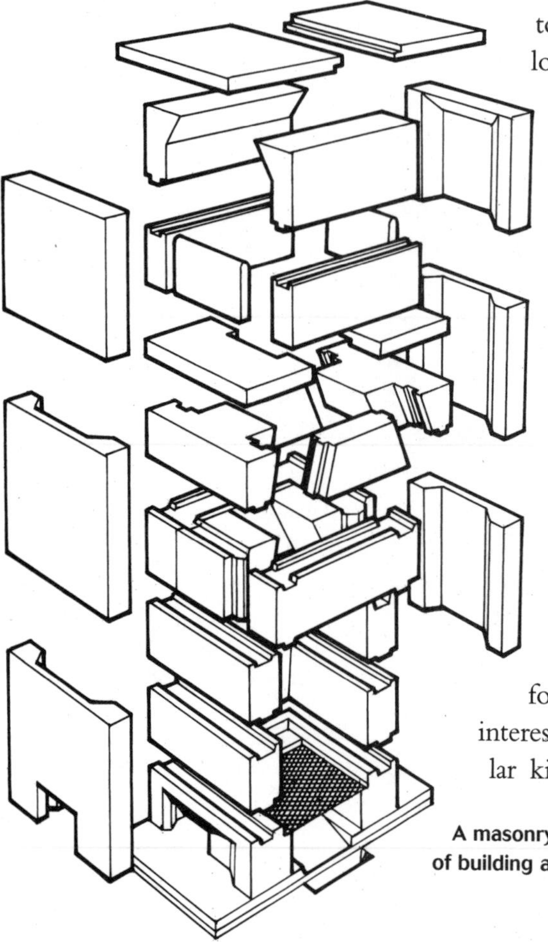

A masonry heater core kit makes the job of building a masonry heater a little easier.

Kits

Building a traditional masonry heater from scratch is a slow, complex process that requires specialized information and training. In Europe, highly skilled masons have been building these heaters for generations, and they continue to construct many thousands of heaters every year. But in the United States and Canada, the limited number of experienced masonry heater artisans has been a problem. In partial response to that difficulty, foreign and domestic companies have created an interesting array of masonry heater kits. These modular kits provide all the basic internal components (often made from a castable refractory material) for a variety of masonry heater designs, as well as detailed instructions on how to build them. The kits are normally shipped to construction sites, where a competent mason assembles them in considerably less time than would be required for a traditional, built-from-bricks design.

Tulikivi, a kit manufacturer from Finland, offers a wide range of soapstone products from fireplaces to cookstoves. An Austrian company, BioFire, manufactures a modular version of the traditional "Kachelöfen." Both of these companies have distributors in the United States. Temp-Cast, a Canadian manufacturer, sells a prefabricated core based on a traditional Finnish design. Masonry Stove Builders, another Canadian manufacturer, offers a modular fireplace heater called a Heat Kit. Check out the MHA Web site (www.mha-net.org) for a complete listing of kit manufacturers and masonry-heater builders.

Safety

A properly constructed masonry heater is one of the safest wood-fired heating devices you can have in your home. You only build a fire in a masonry heater for a few hours once or twice a day, and generally you do not leave a fire burning while you are asleep. After the fire has burned out, there are no exterior parts of the heater that are unsafe to touch, since the surface temperature generally is from 155 to 175 degrees Fahrenheit. You can even sit with your back resting directly against the heater's exterior—something you should never try with a woodstove.

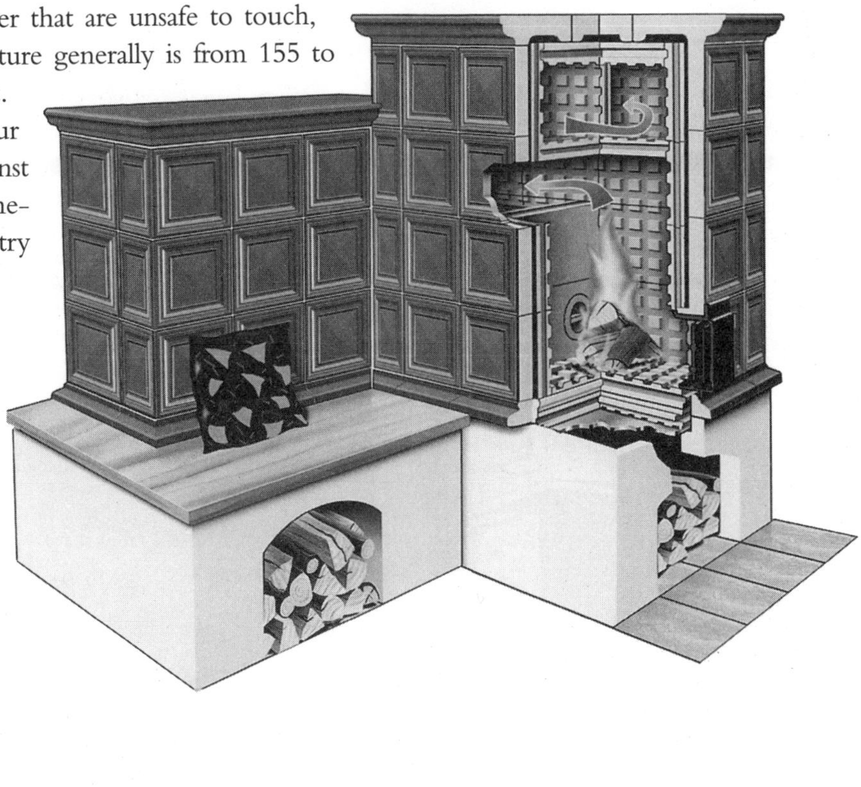

Cutaway view of a masonry heater kit

Courtesy of Biofire, Inc.

Because of their high combustion temperatures, masonry heaters produce almost no particulate emissions and no creosote. A chimney connected to a masonry heater will probably never need to be cleaned due to creosote buildup (but you should check it annually anyway, just to be sure). The lack of creosote formation almost completely precludes the possibility of a chimney fire.

Environmental Issues

Homes using a radiant masonry heater benefit from a gentle, cozy warmth and general freshness of the living-space air. Because a masonry heater does not superheat the air or use fans or blowers, there is virtually no turbulence, draft, or "wind chill." Steel and cast-iron woodstoves and many central hot-air heating systems, on the other hand, tend to burn tiny dust particles in the air, which are then circulated by convection throughout your home, causing an unpleasant burned smell and extremely dry air due to low humidity. This phenomenon can cause respiratory irritation. These problems are eliminated with a masonry heater, which provides steady, even heat throughout your home. And because masonry heaters produce almost no particulate emissions, they have little or no negative impact on the environment outside your home and can be used in most states or localities that restrict other types of wood-burning appliances.

Solar Compatibility

A masonry heater is a perfect match for a solar home, since both are based on passive, nonpolluting renewable energy. If located properly, a masonry heater can add to the heat-storage mass of your passive solar home while acting as your backup heater at the same time (see chapter 5 for information about passive solar homes).

Choosing Your Heater

The ultimate design and appearance of your heater is limited only by your imagination and the abilities of your mason. To help narrow your choices

A custom-designed masonry heater can be a work of art.

Courtesy of Stuart Davies

among the wide range of styles and types of masonry heaters available, start by thinking specifically about your needs. Are you hoping to heat your entire home? In this case, a large, centrally located masonry heater (in an open-style home) would probably be your best choice. Would you like your heater to perform a variety of functions? Which ones? Masonry heaters can be designed with heated benches, cooking surfaces, baking or warming ovens, or domestic-hot-water heating coils. Some masonry heater kits even come in see-through models or can be stacked on top of each other for two-story heating installations.

Installation

The three most important factors about installation of a masonry heater are location, location, and location. A freestanding masonry heater located in the center of a room (or your home) offers the best overall heating potential. Locate the heater so it is visible from as many parts of your living area as possible. That's not only so you can see the heater but so it can "see" you as well. Remember, a masonry heater warms primarily by radiation, so you don't want to block radiant heat by positioning the unit in an obscure corner. Placing the heater against a wall reduces its radiant heating capacity by roughly one quarter; locating it against an exterior wall is even worse. If the heater will serve as a divider between a living room and kitchen area, place the "fireplace" side in the living room and the "baking oven" side in the kitchen.

Operation

A masonry heater is fairly simple to operate, but the daily firing routine is quite different from that for a woodstove. However, as with almost any wood-fired heating appliance, you do need to be mindful of outdoor weather conditions. You build a fire in the firebox once or twice a day, depending on the weather and the size and heat-retention capacity of the heater. If it's extremely cold outside, you use a little more wood. If it's not so cold outside, you burn a little less.

After a few months of learning, you should be able to anticipate how and when to adjust your routine in response to changing weather conditions. Just remember that there is a considerable lag between the time you build the fire and the time the masonry begins to radiate heat. Once the fire has *completely* burned itself out, close the damper.

The same fuel prohibitions for most other wood-fired heating appliances apply to masonry heaters. Follow the operating instructions in your owner's manual if your heater was built from a kit. If a mason constructed your heater from scratch, follow your mason's operating instructions.

Maintenance

Beyond a regular chimney inspection or rare chimney cleaning, maintenance of a masonry heater is about as easy as it gets. You will need to clean the ashes out of the firebox every few days (some masonry heaters have a built-in ash dump located in the basement). Once a year, during the off-season, vacuum the interior parts of your heater that you can access. While you are at it, check for any small cracks or other minor damage. Your mason can make any needed repairs. The gasket on the door is the only part that may need to be replaced occasionally.

Backup

If you are going to use a masonry heater as your primary source of heat, you will need a backup heating system for times when you will be away from your home for more than two or three days during the winter.

Warranties

The warranty on a built-from-bricks masonry heater is a matter between you and your mason. Masonry heater kits offer a variety of limited warranties. Some cast-iron parts, such as doors and glass, generally have a different (usually shorter) warranty period from the modular refractory core parts. If the core fails, however, be aware that the warranty may cover only the cost of the core, not the cost of removal or replacement of façades or chimneys.

Masonry Heater Pros and Cons

Pros

Wood is relatively low in cost and, when burned in a masonry heater, is environmentally neutral.

Masonry heaters are very efficient and produce extremely low emissions.

Masonry heaters can be used in many localities where other stoves and fireplaces are banned or restricted.

Masonry heaters can be used in new construction, while masonry heater kits are better for renovations.

Masonry heaters can be designed to fit a home.

Masonry heaters, if properly located, can also be used as a heat-storage mass in a passive solar home.

Masonry heaters only need to be fired once or twice a day.

Masonry heaters have low operating costs.

Masonry heaters do not require electricity to operate and are a reliable source of heat during a power outage.

The gentle radiant heat from a masonry heater is extremely comfortable.

Masonry heaters are safer than woodstoves for households with small children.

Sitting in front of a masonry heater used as a fireplace is very romantic.

Cons

Good firewood is not available in all locations.

You need a place to store firewood.

Burning wood is labor intensive.

Masonry heaters have high initial capital costs.

Masonry heaters are heavy and require adequate footings and foundations.

Masonry heaters generally need to be built by an experienced mason.

Masonry heaters cannot provide quick heat for your home from a "cold start."

Masonry heaters need to be refueled every day.

Masonry heaters tend to be somewhat dirty to operate.

Where and What to Buy

Although some stove specialty shops may carry (or be able to order) masonry stove kits, in many cases you'll probably have to locate a distributor of kits yourself. The MHA Web site (www.mha-net.org) can be a good resource; the Web site may also help you locate a skilled heater mason to build a heater for you.

If you think a masonry heater makes sense for your situation, and you are building a new home anyway, you can choose from either a built-from-bricks heater or a kit. If you are contemplating a renovation project in an existing house, choosing a masonry heater kit makes more sense. In either case, a kit takes less time to build. Remember, whatever you buy can be customized to almost any exterior appearance you desire (or can afford).

Price Range

Prices for masonry heaters range from $8,000 to over $15,000. An average built-from-bricks model will probably cost around $12,000 ($4,000 for materials and $8,000 for labor, including footings, foundation, and chimney). An average kit will also run about $12,000 ($6,000 for materials and $6,000 for labor, including footings, foundations, and chimney).

PART FOUR

HEATING WITH BIOMASS

CHAPTER 15

Biomass Basics

While burning wood is popular with many folks, it's not for everyone. Perhaps you support the idea of burning wood from an environmental standpoint, but you may not want to commit to the physical labor and the mess associated with a wood-fired appliance. It's also possible that you live in an area where firewood is extremely expensive and hard to obtain or where some wood-burning appliances are restricted or banned. Or maybe you work away from home and can't tend a woodstove every four or five hours. Nevertheless, you still want to do something about our overdependence on fossil fuels, but what?

If you can't burn wood but still want to avoid overdependence on fossil fuels, you can install a stove, furnace, or boiler that burns wood pellets or other biomass fuels. Wood, wood pellets, corn, and other grains are all **biomass**—plant material that can be used as fuel. However, from an operational standpoint, burning other types of biomass is quite different from burning cordwood.

History of Corn and Pellet Heating

Burning corn as a heating fuel began during the Great Depression. Some Midwestern farmers adapted their stoves and furnaces to run on corn instead of coal or wood. They quickly discovered that corn burned cleanly and produced steady, inexpensive heat with substantially less labor and mess than wood. Since then, corn's popularity as a fuel has risen and fallen inversely with respect to the price of corn as a feedstock. Low prices for corn (and high prices for oil and natural gas) in recent years have once again boosted corn's popularity for the home heating market.

The wood pellet market developed more recently, after the Organization of Petroleum Exporting Countries (OPEC) oil embargo of 1973. While the scarcity and high price of heating oil was a key factor, it wasn't the only one. For years, the U.S. Environmental Protection Agency (EPA) had been trying to find uses for the sawdust that was routinely being dumped by most sawmills across the country. That environmental initiative, combined with the oil scare and subsequent price spikes in the heating oil market, resulted in the development of the first residential wood **pellet stove** in 1983. These stoves, introduced to the market the following year, offered new levels of automation and convenience for heating with wood. Like most new machinery, the early models initially suffered from mechanical difficulties, but by 1990 there were more than fifteen manufacturers of improved-design wood pellet stoves in this country. The sales of pellet stoves increased rapidly in the early 1990s but, after reaching a peak in 1994, began to decline as competition from gas-fired stoves increased dramatically. Nevertheless, by 1996, the number of pellet-stove manufacturers had risen to twenty, and about eighty mills were producing high-quality pellets. Pellet stove sales rebounded again after the oil and gas price spike in 2000 and 2001 and have remained strong ever since.

In the United States today, residential use of pellets accounts for about 95 percent of sales; industrial or commercial use makes up the balance. Between 1993 and 1998, pellet fuel sales amounted to between 500,000 and 600,000 tons annually. That figure is now around 750,000 tons for the approximately 500,000 pellet stoves and fireplace inserts in homes throughout the United States and Canada, according to the Pellet Fuels Institute (PFI) in Arlington, Virginia. But to keep all of this in perspective, it's important to understand that pellet fuel still only accounts for a mere 0.025 percent of residential space heating in the United States.

Pellets

"The main advantage of pellets is that they are a lot more convenient than wood; you just need to use one bag of pellets a day on average," says Steve Walker, the founder and president of New England Wood Pellets, located in Jaffrey, New Hampshire. "Pellet-fired appliances are thermostatically controlled and automatically fed so you don't have to tend them much. As long as you can pour in the bag of pellets and push the 'On' button, that's about all there is to it. A pellet heater is easier, simpler, and more conducive to the modern-day work schedule, with both adult family members working. As long as someone is home at least once a day, you're all set."

Walker cites other advantages to pellet heat as well. "No matter how you burn it, biomass is far better as far as global warming and carbon dioxide emissions are concerned. Pellets don't contribute any net carbon dioxide because the trees grown as replacement stock absorb the carbon dioxide resulting from burning the pellets. Relative to woodstoves, there are far fewer particulate emissions and smoke, to the point where pellet stoves don't need to be vented up a standard chimney, which makes the total installation cost more competitive with a woodstove." Best of all, pellet-fired appliances are approved for use in virtually all regions where other wood heaters may be restricted or banned. This makes pellet-fired heating appliances attractive not only as primary heat sources, but also as secondary sources for people who want to maintain some control over their heating expenses in an increasingly volatile market.

Although they look a lot like animal feed, pellets are manufactured from biomass feedstocks such as trees and plants. The pellets most commonly used for residential heating in the United States are made from sawdust or ground wood chips, which are waste materials from the manufacture of lumber, furniture, and other forest products. The naturally occurring resins and binders in the sawdust allow the pellets to maintain their shape without the need for any additives. Pellets can also be made from agricultural waste products such as corn stalks and nut hulls or from crops such as switchgrass.

Pellets are easier to handle than firewood. Pellets flow fairly easily through hoppers and augers, and they take up less space relative to the amount of heat produced than wood. But, in the United States, pellets are not as easy to use as oil or gas, since delivery, storage, and burning of these fossil fuels requires no effort on the part of the homeowner (except to pay the fuel bill). In Sweden, however, many people routinely receive bulk delivery of wood pellets by a delivery truck to a storage bin in their homes. The pellets

Pellet Progress

Sweden didn't begin to produce and use pellets for fuel until the early 1990s. However, by 1996, Sweden's annual consumption of pellet fuel surpassed that of the United States.

are then augered from the storage bin directly into a thermostatically controlled central heating boiler or furnace in the basement. This sophisticated approach makes heating with pellets virtually as convenient as heating with oil or gas. The first bulk delivery truck for wood pellets in the United States began making deliveries in parts of New England in 2002 as part of a demonstration project.

Pellet Manufacturing

Pellets are made at mills specifically designed for their production. Some of these mills are stand-alone facilities, while others are affiliated with or part of wood-products manufacturing plants. The first pellet mill in North America was built in the 1970s in Oregon to supply wood pellet fuel for commercial and industrial uses. Pellet mills obtain, sort, grind, dry, compress, and bag the pellets, which can be made from either softwood or hardwood sawdust. Once bagged, the pellets are shipped to local retailers, including stove and fireplace specialty stores, hardware stores, garden supply stores, feed mills, and other outlets.

Characteristics

Because the chemical makeup and moisture content of various biomass ingredients can vary, the PFI has developed voluntary fuel standards. These standards cover such characteristics as density, size, amount of fines (dust), salt content, and ash content. Pellets can be standard grade or premium grade; the main difference between the two grades is ash content. Standard-grade pellets (up to 3 percent ash content) are usually made from materials that result in more residual ash when burned; examples include sawdust that contains bark or agricultural residues such as nut hulls. The higher ash content is due mainly to impurities contained in the bark or hulls. Ash content is important because it plays a key role in pellet appliance maintenance; the greater the ash, the more frequent the maintenance. Standard-grade pellets should only be used in heating appliances that are specifically designed for them.

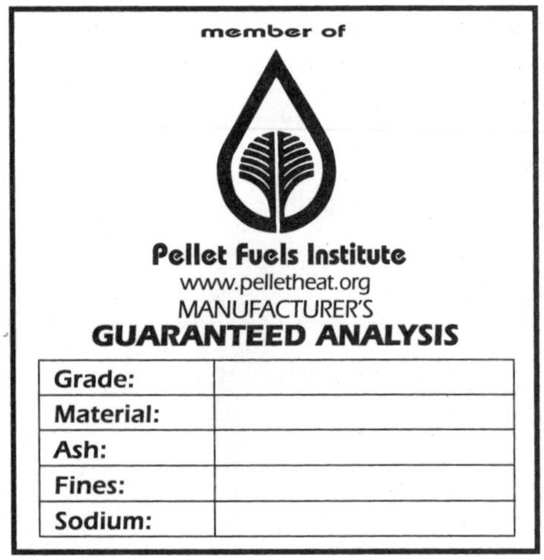

member of

Pellet Fuels Institute
www.pelletheat.org
MANUFACTURER'S
GUARANTEED ANALYSIS

Grade:	
Material:	
Ash:	
Fines:	
Sodium:	

The Pellet Fuels Institute (PFI) guaranteed-analysis label offers information on the density, size, amount of fines (dust), salt, and ash content of wood pellets.

Courtesy of Pellet Fuels Institute

Premium-grade fuel pellets contain less than 1 percent ash and are usually made from sawdust that does not contain any bark. Premium-grade pellets can generally be used in stoves manufactured to burn either standard or premium pellets. Premium pellets are far more popular, accounting for about 95 percent of the present pellet market.

One characteristic of pellet fuels not specifically included in the PFI standards is trace mineral content. Some minerals such as silica can promote clinkering, the formation of fused ash that can interfere with the combustion process (see "Burner Design" on page 185). All pellets have a low moisture content (below 10 percent), which contributes to high combustion efficiency and extremely low emissions. For all practical purposes, pellet appliances do not produce creosote.

Handling and Storage

Working with pellets is much easier than dealing with cordwood. A ton of pellets can be stored in an area about one-third the size required for a cord of wood. The easily handled, 40-pound bags of pellets can be stacked in a dry garage, basement, utility room, or storage shed. And because bags of pellets are clean, they can be stored and handled in your home without creating the usual mess associated with firewood. Best of all, there's no cutting, splitting, and drying.

Bags of pellets are clean and easy to handle

Courtesy of Amber Rood and Greg Pahl

Costs

As with most other fuels, the price of pellets varies from one region to another and from season to season. Since the costs of production are about the same for most pellet mills, the main reason for the price variation is the cost of the sawdust. In locations such as British Columbia, Canada, sawdust is very inexpensive because of the enormous supply. In New England, at the other end of the spectrum, sawdust is no longer dumped and the increased demand has driven up the cost of sawdust to almost $50 per ton for some pellet mills. Pellets are sold for between $120 and $200 per ton, with an average of about $150 per ton. You can buy a single 40-pound bag of pellets for $3 to $4. There are 50 bags per palletized ton (some mills offer

20-pound bags for easier handling). A ton of pellets has the heat value of about 1½ cords of wood. The average home will need about 2½ tons of pellets for the heating season. The best strategy is to stock up on your winter's supply of pellets during the summer, while prices are lowest.

Environmental Issues

From an environmental standpoint, burning wood pellets offers all positives. Pellets are a renewable resource, so burning pellets does not add any net carbon dioxide to the atmosphere. Because of the extremely hot combustion in most pellet-fired appliances, pellets burn cleanly with extremely low emissions and produce virtually no creosote. Compared to other fuels, burning 1 ton of pellets instead of heating with electricity will save 3,323 pounds of carbon emissions. You'll save 943 pounds of emissions per ton by replacing oil with pellets and 549 pounds per ton if you're replacing natural gas. Most pellets are made from sawdust, a by-product of the forest products industry that was formerly dumped or burned.

Pellet Pros and Cons

Pros

- Pellets are cost competitive with many other fuels.
- Bags of pellets are easy to handle and stack compactly.
- Pellets are clean burning.
- Pellets have a positive environmental impact.
- Pellets offer steady heat.

Cons

- Pellets require a special heating appliance.
- Pellet heating appliances (except some furnaces and boilers) must be tended about once a day.
- Pellets require storage space and some (light) physical labor.

Corn and Other Biomass Fuels

The second most popular biomass fuel is corn. When most people think of corn, they generally think of eating it. Hot, buttered corn-on-the-cob is hard to beat, especially when it is freshly picked and cooked. Of course, corn is also used as an ingredient in countless foods and other products and as a feed for livestock. Nevertheless, corn is a viable fuel under the right circumstances.

But wait a minute! If you try to use corn as fuel in a heating appliance, won't you suddenly be swamped with enough popcorn to feed the whole neighborhood? That's a logical question, and the answer is no. Popcorn is a specific variety of corn. Dried corn grown for animal feed will not pop in your stove or furnace. Some people, however, say that a home heated with corn often exudes a faint, pleasant scent of cornbread baking in the oven. That sure beats the smell of conventional home heating oil.

To be an effective heating fuel, shelled corn must be dry. Attempting to burn wet corn is just as frustrating and nonproductive as trying to burn wet wood, and I don't recommend it (neither do corn-fired heating appliance manufacturers). A pound of dry (15 percent moisture) corn will produce about 7,000 Btu, versus about 8,000 Btu for wood and 9,000 Btu for pellets (a ton of corn produces 14 million Btu). Although the heat output for dry corn is somewhat less than wood, corn is easier to handle than firewood because it flows fairly well through hoppers and augers and has a better energy-to-volume ratio. Using corn as a fuel requires a special corn-burning appliance (although some pellet appliances will burn corn as well). In order to flow properly through these appliances, however, corn must be clean and free of cob pieces or excessive fines (corn dust). The quality of the corn is not that important, as long as it is clean and dry. In fact, it's perfectly acceptable (or even preferable) to burn lower-grade corn that is not fit for the feed market. Lower-grade corn is usually cheaper, sometimes even free, if you can locate a local farmer who wants to get rid of a bad batch.

A popcorn appliance?

Availability

Corn is available all across the United States and Canada but is generally cheapest in the Midwest, where the supply is enormous. You can buy corn at feed stores and mills, farmer's markets, and perhaps in the gardening sections of some large retail outlets. Farmers in your local area are probably the best source for the lowest prices. You may have to haul the corn yourself, however.

Storage

Storing corn in your home can be problematic, because rodents love corn. Stored corn can also attract grain insects and moisture. Most farmers generally have sophisticated (and expensive) grain storage silos or bins that are designed to minimize the pest problem. These storage facilities are simply part of the cost of doing business for farmers. The fact that these facilities are a handy place to store corn for heating purposes is coincidental. Most homeowners don't have the luxury of ready-made grain storage facilities. If you want to burn corn for heat, you'll have to figure out how to protect it (and your home) from mice, rats, and insects. You can't simply stack bags of corn in your garage or basement as you would bags of wood pellets. Large metal garbage cans with tightly fitted lids will work for relatively small quantities of corn, but trying to figure out how to store several tons of corn is a real puzzle. At the very least, you will have to try to construct or buy some kind of large, rodent-proof storage bin. This problem usually discourages all but the most enthusiastic homeowner from choosing to burn corn for heating.

A Nifty Fuel-Choice Chart

Check out the energy-selector Web site for corn hosted by Penn State University (http://energy.cas.psu.edu/energyselector/), which shows the fuel switching point between corn and most other fuels. Simply choose the fuel you wish to compare, and the resulting chart will tell you which fuel to burn under a wide variety of price scenarios. It's easy. And best of all, no formulas!

Costs

Corn is generally sold by the bushel, which costs between $2 and $4 (to help nonfarmers visualize this better, a bushel of corn will roughly fill two 5-gallon buckets). For the sake of comparison, the heating value of 63 bushels of shelled corn is roughly equivalent to one cord of firewood. You will get the best price on corn, however, if you buy it by the ton. But because the cost of corn tends to fluctuate substantially over time, it may or may not be a good choice as a heating fuel.

Other Biomass Fuels

A wide variety of other grains, including wheat, barley, rye, sorghum, and soybeans, are also generally suitable for use as heating fuels. The same considerations that apply to corn also apply to these grains. Burning grain for heating is a strategy that is mainly attractive to farmers who already have the facilities to store and handle grain. For most people who live in cities, burning grains for heating purposes is not feasible. If you aren't a farmer and want to try this strategy, by all means, go ahead. Just be aware of the drawbacks before you invest a lot of time, effort, and money in the project.

There are also agricultural waste products such as wheat straw, corn stover (dried cornstalks), and orchard trimmings that can be burned successfully for home heating. However, as a practical matter, these fuels are limited mainly to the farming sector, except when the waste material is turned into pellets.

Environmental Issues

From an environmental standpoint, burning corn offers some advantages but is not entirely benign. Corn is a renewable resource that does not add any net carbon dioxide to the atmosphere. Because of the extremely hot combustion in most corn-fired appliances, corn burns cleanly, with extremely low emissions.

However, the heavy use of fertilizers, pesticides, and fuel to produce and transport corn is an environmental negative. The production of corn also tends to result in more soil erosion than does the production of some other field crops. One could also be tempted to ask what we are doing burning feed grains for fuel when there are so many hungry people on the planet. This raises a host of difficult and troubling economic, political, and social questions that extend beyond the scope of this book. Think about it, and then decide for yourself.

Grain Pros and Cons

Pros	Cons
Very competitive costs when grain prices are low	Requires a special heating appliance
Relatively easy to handle	Heating appliance often must be tended about once a day
Clean burning	Secure storage space and some physical labor required
Steady heat	Stored corn and other grains attract rodents and insects
Relatively low environmental impact	

CHAPTER 16

Pellet Stoves

During the energy crisis of 1973, the Boeing Company asked Dr. Jerry Whitfield, a fuel efficiency engineer, to develop alternative fuels for industry. The idea was that if industry could switch to other fuels, then more of the scarce petroleum-based fuel could be reserved for aircraft. As Whitfield embarked on his research, he could not have foreseen that he was about to become a pioneer in the pellet stove industry.

Whitfield, who had a background in fluid-bed (forced-air) furnace technology, eventually met Ken Tucker in Idaho. (Tucker is now the president and CEO of Lignetics, Inc., a leader in the wood pellet fuel industry.) Tucker was intrigued by the way alfalfa could be turned into pelletized animal feed. He hypothesized that the same general process could be used to manufacture pellet fuel from sawdust and that the pellets could be burned in industrial furnaces. But Tucker's experiments had not been successful until Whitfield arrived on the scene.

Whitfield was impressed with Tucker's work and quickly realized that the heating potential of pellet fuel extended well beyond industrial applications. Whitfield applied his knowledge of forced-air furnace technology to smaller-scale residential heating systems, and in 1983, he introduced the first forced-air pellet stoves. Slow to catch on at first, pellet stoves were selling at about seventy-two thousand units annually when they hit their peak in 1994. Despite some variability in the market, pellet stove sales remain strong. More and more people are discovering the many benefits of these remarkable heating appliances.

How They Work

Unlike its cordwood relatives, a pellet stove limits much of the physical labor of operation through automation. The best part of this automated design is

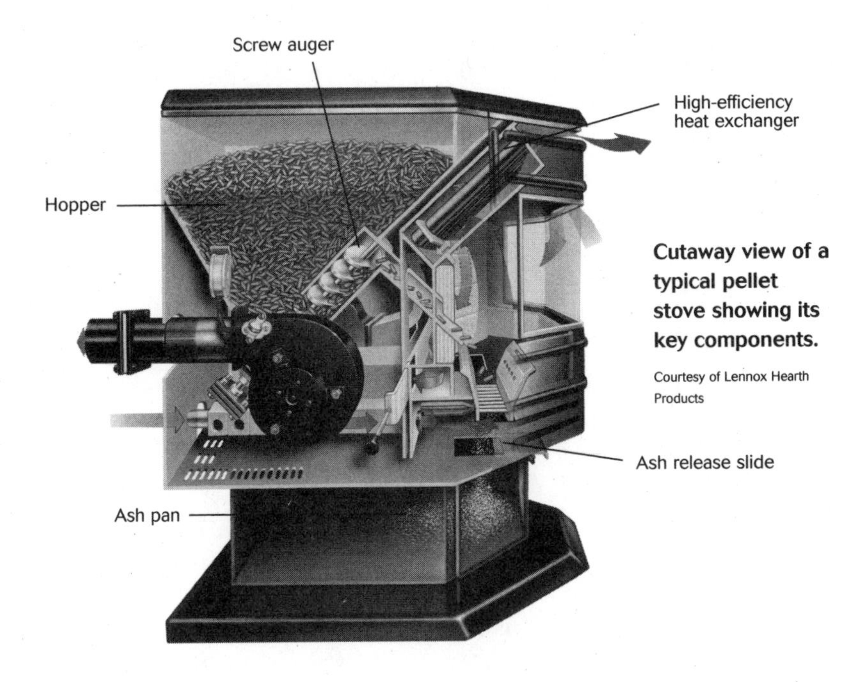

Cutaway view of a typical pellet stove showing its key components.

Courtesy of Lennox Hearth Products

in the fueling process. An auger feeds pellets to the fire in carefully controlled amounts; it draws the pellets from a storage bin or hopper (which is usually built into the main part of the stove). The speed at which the auger turns controls the amount of fuel added and the amount of heat produced. Not only does this technology eliminate the chore of hand-fueling, but also it allows pellet stoves to generate just the right amount of heat to maintain a constant temperature, which can be regulated by a thermostat. This ability to produce steady heat is generally not possible with most other heating appliances. Even most oil- and gas-fired central furnaces cycle on and off, allowing for considerable fluctuations in temperature between firings. Most pellet stoves eliminate these annoying fluctuations.

Better yet, most pellet stoves only need to have their storage hoppers filled once a day, making these heaters an excellent choice for people who want to burn wood but who are away from home most of the day. The frequency of refueling does depend on the size of the storage hopper and on the weather. The pellets are combusted in a burn pot (a small combustion chamber) with the aid of a blower, creating a mini-furnace, which generates an extremely hot and efficient fire with very low emissions. The combustion efficiency of many pellet stoves is around 95 percent. They produce lower emissions than any other type of solid-fuel-burning hearth appliance and are an obvious choice in locations where air quality in the winter is an issue.

The blower, which can have more than one speed setting, also directs the combustion gases out of the stove and into a small vent pipe instead of a normal chimney. Before exiting the stove, the heat from the fire normally passes through a high-efficiency heat exchanger, which warms the air in your living space, either by convection or with the assistance of a circulating blower. The heat exchanger and forced combustion-air design allows pellet stoves to achieve overall heating efficiencies between 60 and 80 percent.

Types

Pellet stoves are available in a wide variety of styles, sizes, and finishes. The three main categories of these stoves are freestanding stoves, fireplace inserts, and built-in appliances.

Freestanding

The most popular pellet stove is the freestanding design. Legs or a pedestal supports this type of stove, which offers the most flexibility in terms of installation; it can be located almost anywhere in the living space of your home. Although it must be placed on a noncombustible floor protector, a freestanding pellet stove can usually be installed much closer to walls and other combustible surfaces than a typical cordwood stove.

Fireplace Inserts

Like their cordwood-burning cousins, pellet-burning fireplace inserts can be installed in most functional masonry fireplaces. A metal shield (or shroud) seals off the rest of the opening between the insert and the fireplace mouth. Some pellet inserts are approved only for use in masonry fireplaces, while others can also be installed in approved factory-built metal fireplaces.

A freestanding pellet stove is the most popular design.

Courtesy of Quadra-Fire

Built-In Appliances

A pellet-fired built-in (or zero-clearance) appliance offers the warm ambiance of a fireplace without much of the expense of a traditional masonry fireplace. Although a noncombustible floor protector is required, these units can be framed in with fairly tight clearances to combustible mate-

rials. When faced with brick, tile, or stone, a built-in appliance looks similar to a traditional fireplace. Many pellet fireplace inserts have zero-clearance kit options.

A pellet fireplace insert transforms an inefficient fireplace into a serious heating source.

Courtesy of Quadra-Fire

Burner Design

Pellet appliances are further subdivided into two additional categories based on their burner design: top feed and bottom feed. **Top-feed** stoves have an auger that delivers the pellets to a tube or chute above the burn pot. In this design, the pellets simply fall into the burn pot from above. In a **bottom-feed** stove, the pellets are augered to the fire from below or behind the burn pot. Although there are variations, the bottom-feed design generally allows for more flexibility in the grade of pellet that can be burned, because the feeding action of the pellets pushes the ash and **clinkers** (fused minerals) away from the burn area. This self-cleaning action of the bottom-feed design reduces the amount of manual cleaning required. Some top-feed stoves have specially designed grates that allow heavier ash and clinkers to fall through to the ash pan; other are designed with a burn pot that rotates to help keep the air inlets open. Grates that can be cleaned without interrupting the operation of the stove are especially convenient. Although there are some exceptions, top-feed designs are common in pellet stoves, while bottom-feed designs are more common in pellet furnaces and boilers (see chapter 17 for details on these).

Convenience Features

One important feature to consider when comparing pellet stoves and pellet fireplace inserts is the pellet capacity of the storage hopper. Capacity varies from 40 to over 100 pounds in some models. The quantity of pellets you can load into the hopper at one time is a key factor in determining the frequency of refueling. Ideally, you shouldn't have to refuel more than once a day. Another important feature is whether your pellet stove has manual or automatic controls. Some people prefer to just "set it and forget it," while others enjoy a more hands-on relationship with their heating appliances. If cost is a major issue, note that manually controlled pellet stoves are less expensive than their automatic counterparts.

Pellet stoves use three types of ignition systems: standard, self-starting, and fully automatic. Standard ignition requires the use of a starter gel and a match. Self-starting ignitions employ a button or remote control to light the fire. Fully automatic ignition is controlled by a thermostat and allows the stove to cycle on and off, depending on the heat level selected. Automatic ignition, available on many newer-model pellet stoves, may be more useful on stoves that are used intermittently rather than continuously. Automatic-ignition pellet stoves can switch between high and low settings and turn themselves off when heat is not required; standard and self-starting stoves can switch between settings but cannot shut themselves off. Remote thermostats are also available for many pellet stoves.

Other Features

One of the nicest features offered on both freestanding stoves and fireplace inserts is your ability to watch the fire through an airtight glass door (the firebox door should be kept closed while the stove is operating). The glass is kept clean by the same type of air-wash design that is used on many woodstoves. Watching the fire in a pellet stove or insert offers an ambiance similar to that of sitting in front of a woodstove or fireplace. However, watching the flames from a superheated pellet fire in an otherwise blank scene is underwhelming for those who are used to watching a real cordwood fire.

Most pellet-stove manufacturers realize this and offer a variety of imitation log sets that can be placed in the firebox area. Some of these log sets can be reasonably convincing with flames dancing up through them. Others look obviously artificial. Regardless of their appearance, the best feature of these logs is that they don't ever have to be replenished. That's something that a cordwood stove can't duplicate. Find out if the log set for your stove is easily removable during cleaning procedures; if possible, avoid those that aren't. Other optional features include self-cleaning burn pots, oversize pedestal ash pans, gold or nickel door trim, and special porcelain or cast-iron finishes.

Although pellet stoves produce some direct radiational heating through the glass window in the door, the majority of the heat is generated by convection, usually with the aid of an internal fan or blower. This means that the exterior surfaces of most pellet-stove cabinets tend to stay relatively cool, reducing the chance of accidental burns from touching them. But the fans can be noisy. While you are at a stove specialty shop, listen carefully to the blower on the model you are interested in while it is operating. Some models are quieter than others. Blower noise can be especially important to consider if the stove or insert is installed in a prominent location in your living area.

Sizing a Stove

A pellet stove always burns at high efficiency regardless of the thermostat setting. Because the heat output of pellet stoves is easily controllable, sizing is not quite as crucial as it is with a cordwood stove. Nevertheless, most pellet-stove manufacturers offer several sizes, and it is important to pick the heat output that best matches your home's heating requirements and your weather zone, as well as your expectations and needs. Remember that a stove, regardless of its fuel, is primarily a heating appliance for a single room or zone. Because of their hot-air circulating fans, however, most pellet stoves do a better job of moving warm air through a home than a traditional woodstove.

Installation

The blower-assisted venting feature on pellet stoves allows them to be installed almost anywhere, since a standard masonry chimney is not required. However, one of the necessities for the installation of a pellet stove or insert is a nearby electrical outlet. If an outlet is not available, you will need to have one installed. Because of their design, pellet stoves require less clearance space than most other types of stoves and, in some cases, can be located as little as three inches from a wall (follow the clearance guidelines in your stove's installation instructions, as well as all applicable building codes for your location). If you are lucky enough to have an existing hearth where you plan to install a pellet stove, you probably won't even have to worry about floor protection. If not, you will need to use a noncombustible floor protector sized according to the installation instructions for your stove.

Like any heating appliance that uses combustion to produce heat, pellet stoves and inserts must be properly vented. Although pellet stoves produce almost no visible smoke after they have started up, the exhaust gases need to be removed safely from the appliance. For most pellet stoves, the exhaust process is facilitated by the blower that expels combustion gases through the vent pipe while simultaneously drawing in fresh combustion air. A specially manufactured 3- or 4-inch-diameter "L" or "PL" vent pipe with a double-wall design is required for most pellet stoves. Since 4-inch pipe is not much more expensive than 3-inch pipe, some pellet stove retailers suggest using the larger diameter pipe, which reduces potential maintenance headaches. Follow the instructions that come with your stove on the method to

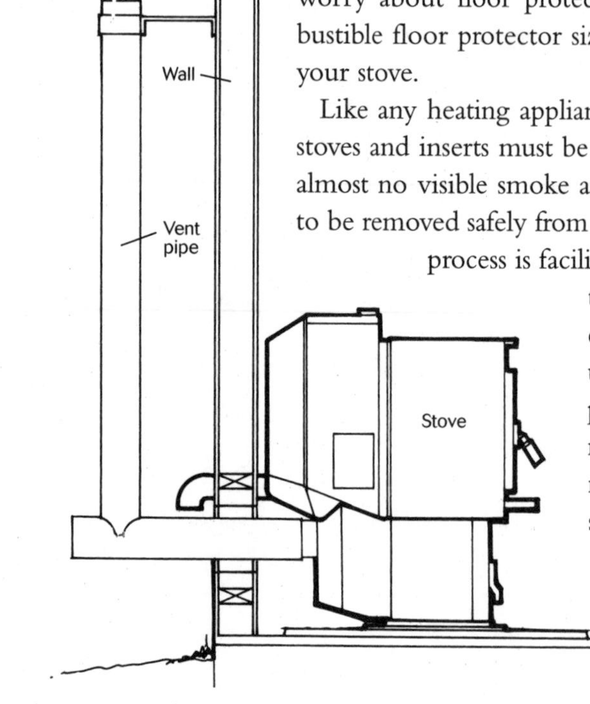

A simple through-the-wall vent pipe installation.

secure and seal the joints of this vent piping. Do not use *any* of the following materials to vent a pellet stove:

Dryer vent
Gas appliance (Type B) vent
PVC (plastic) pipe
Single-wall stove pipe (unless specifically approved by your stove's installation manual and local codes)

PL or L vent pipe can be installed through a side wall, a ceiling and roof, or into an existing chimney. Have an existing chimney inspected to ensure that it is clean and structurally sound before installing vent pipe into it. Some pellet-stove installations may require that a masonry chimney be partially or completely relined.

Operation

The operation of a pellet stove is fairly simple, but if the technology is unfamiliar to you, be especially careful to read and follow the operating instructions that come with your unit. The usual routine is to add a bag of pellets to the built-in storage hopper, set the controls for start-up phase, press the start button, and sit back and enjoy the fire.

In the case of a manual-ignition stove, you will have to apply a starter gel or other approved solid starter material, light the pellets, and watch the fire to be sure it catches properly (never use gasoline or other flammable liquids to start the fire). Once the fire is burning briskly, an air-inlet adjustment (on manual models) is about all that is necessary for many hours of even, comfortable heat.

Shutting the stove down is normally accomplished by setting the control to the "Off" position and then waiting for the fire to burn out. Once the stove has cooled sufficiently, all motors and blowers will shut down. Never turn off your pellet stove by unplugging it from the electrical outlet, because

Watch Your Fingers

When filling the hopper of a pellet stove with pellets, always keep your fingers away from the auger. There is potential for serious injury if the auger should unexpectedly start to operate.

this will disrupt the normal shutdown procedure and combustion gases may spill into your living area. During a power failure, if you do not have any backup power, the auger will stop turning and pellets will stop feeding into the burn pot. However, the fire will continue to burn or smolder until the fuel has been consumed. This may take anywhere from a few minutes to an hour or longer. Manual-ignition stoves will not restart automatically after a power outage; you will have to relight the fire.

Depending on how your stove is vented, it is possible that some combustion gases may escape into your home during a power failure unless the vent has at least eight feet of vertical vent pipe to provide for natural draft. In addition, negative pressure caused by a large-capacity exhaust fan in the kitchen or elsewhere can also cause combustion gases from a pellet stove or insert to spill into your living area. These are issues that you should discuss thoroughly with your professional stove installer.

Because they burn so cleanly, pellet stoves produce very little ash. During the winter, most pellet stoves only produce one or two cups of ashes every few days. For many models, you need to empty the ash pan only once every few weeks. On some models, the ash pan is large enough that it has to be emptied only two or three times a year. That's a feature that a traditional cordwood-stove owner can only dream about. Just as with cordwood heating appliances, ashes removed from a pellet stove should be stored in a metal container with a tightly fitted top pending final disposition. Be sure to place the metal container on a noncombustible surface.

Maintenance

Pellet stoves are fairly easy to maintain. Routine tasks include adding pellets to the hopper, emptying the ash pan when needed, and periodically cleaning the burn pot, hopper, ash traps, and door glass. The hopper and auger tube should occasionally be emptied completely to prevent the accumulation of too many sawdust fines. Most manufacturers advise removing unused pellets from the hopper and auger at the end of the heating season to reduce the chance of rusting. You can easily perform all of these routine chores yourself. Depending on the design of your stove, you may also be able to easily clean a heat exchanger that has an external rod or handle. Gaskets on the door, ash pan, and (sometimes) the hopper lid may occasionally need to be replaced.

Because they are fairly sophisticated appliances with electronic controls and moving mechanical parts, pellet stoves should receive a thorough annual servicing. If you enjoy this sort of work (and your warranty allows you to do it yourself) you can follow the annual-maintenance instructions that come with your unit. If not, have the work done professionally. (Finding competent in-home service is something that you should do before you buy a pellet stove or insert.) The cost of the service call will probably be about the same as for any standard oil- or gas-burning furnace or boiler. Most pellet stoves and inserts do not generate appreciable amounts of creosote, but the vent or chimney should be checked annually and cleaned if necessary.

Backup

Virtually all pellet stoves require electricity to operate (about one hundred kilowatt-hours per month). Consequently, if you are going to rely on a pellet stove as your main heat source, you need a backup strategy for electricity during power outages. Many pellet stoves have optional backup battery packs available; these allow the stove to operate for a few hours during a power outage. A good system will automatically switch from utility power to batteries and back again. However, some people (including some pellet-stove dealers) suggest that buying an electrical generator is a better strategy, since you can use the generator to power more than just your stove. You can buy a reasonably capable generator for around $600. Be sure to have a licensed electrician help you set up the system with a properly installed bypass switch.

Warranties

Pellet stoves generally come with limited five-year warranties on the firebox and one- to three-year limited warranties on electrical components. Some manufacturers offer optional warranty extensions for between one and five years. Be sure to read the warranty on your pellet stove carefully, especially the fine print. Never burn waste paper, cardboard, or any material other than approved fuels in a pellet stove; doing so will void your warranty. Some warranties specifically prohibit burning corn or other grains.

Pellet Stove Pros and Cons

Pros	Cons
Pellet stoves and inserts can be used in new construction or renovations.	You need a place to store the pellets.
They provide steady, easily controlled heat.	Most pellet stoves and inserts require electricity to operate.
They only need to be refueled about once a day.	The sound of the blower(s) may bother some people.
They do not require a masonry chimney.	Some light physical labor is required.
The stove exterior (except the glass door) remains relatively cool, reducing the risk of accidental burns.	You will need some form of backup if you are going to be away from home for more than a day.
Sitting in front of a pellet stove or insert is relatively romantic.	Combustion gases may spill into your living space during a power outage.

Where and What to Buy

A stove-and-fireplace specialty shop is the best place to seek information on pellet stoves and pellet inserts (in some cases, you may have to go to a pellet-stove specialty shop). Not only will the employees in a specialty shop be able to provide you with competent advice for your particular situation and needs, but also they will be able to install professionally the unit you decide to buy.

The type of pellet stove you buy is mainly determined by the location where you want to install it, your style preference, and your budget. If you have a wide choice of potential locations in your home, a freestanding model probably makes the most sense. If you have a functional-but-inefficient fireplace that you want to transform into a serious heating appliance, than a pellet insert would be your best choice. If you are on a limited budget but still would like a fireplace experience, a pellet-fired built-in appliance might suit your needs. Whatever type of stove you choose, be sure to look for heavy-duty construction and components. Avoid stoves constructed from thin sheet metal, because they may burn out or rust in just a few years. Some of the most prominent pellet-stove manufacturers are Aladdin Hearth Products, The Earth Stove, Harman Stove Company, Thelin Company, Travis Industries, and Whitfield Hearth Products.

Price Range

The price of pellet stoves runs from $1,500 to $3,000. Add another $400 to $500 for a typical installation. Generally, pellet stoves are somewhat more expensive than cordwood stoves. However, since a pellet stove can be installed without a traditional masonry or Class A chimney, the *total* installed cost of a pellet stove may actually be less than a cordwood stove. Since a pellet stove is a complex device with the potential for mechanical and electrical failures, it makes sense to buy one that is well made. Trying to save a few dollars on a "bargain brand" may end up costing you much more in repairs over the long run.

CHAPTER 17

Pellet Furnaces and Boilers

In 1982, a farmer walked into Traeger Heating in Mt. Angel, Oregon, with a bucket of wood pellets. He asked Joe Traeger, the heating company's owner, to figure out a way to burn the pellets as a heating fuel. Traeger, a heating-business veteran who had developed a wood-fired furnace in 1978, was intrigued. He started experimenting. Traeger first tried using the pellets in an old wood-fired boiler, but they didn't burn very well. Next, he added a blower, which improved combustion substantially. Then, he installed a small burn pot, with even better results. By the end of the year, Traeger had perfected the design enough to start selling a few pellet furnaces and pellet stoves locally.

Over the next five years, sales of Traeger's stoves and furnaces steadily increased, and in 1987, the stove line was licensed to The Earth Stove. Demand for the furnaces continued to expand. In 1992, a boiler was added to the product line. Today, Traeger pellet furnaces and boilers are manufactured and distributed by Pinnacle Stove Sales, Inc. in Quesnel, British Columbia. While there are a few other manufacturers, Traeger offers the largest selection of pellet furnaces and boilers in the United States and Canada. The line ranges from a 70,000 Btu residential model to a 400,000 Btu commercial unit.

Pellet-Fired Central Heat

Sometimes pellet stoves aren't practical as whole-house heaters because of architectural design that restricts the circulation of heat from one room to the next. Pellet furnaces and boilers are an intriguing alternative. These appliances combine most of the positive features of pellet stoves with the advantages of central heating, making them an excellent choice for large homes with numerous separate rooms or additions. Although many people enjoy sitting around a fire in the living room, other people—especially those with young children—prefer to have their heating systems safely down in the basement. If you have a home that would be hard to heat with a stove, or if you already have an existing hot-air or hydronic heating system but want to burn a renewable fuel, you may want to consider a pellet-fired furnace or boiler as a conversion or add-on unit. And if your furnace or boiler is nearing the end of its career, installing a pellet-fired unit might be an excellent replacement strategy.

Pellet-fired central heating systems, like their cordwood-burning counterparts, are normally installed in the basement or a utility room. These furnaces or boilers distribute their heat with a new or existing forced hot-air or circulating hot-water system and can easily heat any size home. There are several factors to consider if you are interested in a pellet-fired heating system. It's important to find a reliable supplier of pellets in your area and to think about whether you are willing to engage in modest amounts of labor to operate and maintain your heating system. The proliferation of pellet mills has largely resolved the pellet availability question. Nevertheless, a pellet-fired central heating system does requires some advance planning for fuel, some light physical labor, and fairly regular attention during the heating season. Compared to a cordwood-burning furnace or boiler, however, pellet-fired units involve much less labor and very little mess.

Pellet-fired central heating represents a tiny niche in the overall heating market in the United States, due primarily to the generally low cost of heating oil and natural gas. But in northern Europe, pellet-fired central heat is increasingly popular. Consequently, Europe (Sweden and Denmark in particular) has been the source of some of the most advanced pellet central heating technologies in recent years. Several Swedish manufacturers even produce a wide range of conversion pellet burners for many models of furnaces and boilers. Unfortunately, these units are not readily available in the United States. Some residential pellet furnace and boiler systems in countries such as Sweden have been automated to the point where there is virtually no labor involved on the part of the homeowner. I am convinced that if these types of sophisticated installations were more common in the United States, pellet furnaces and boilers would become strong contenders for heating our homes. At the moment, however, we're faced with a classic catch-22; pellet furnaces and boilers won't become popular until bulk delivery and storage of pellet fuel is widely available, but bulk delivery of pellet fuel won't become widespread until there is a critical mass of installed pellet-fired central heaters.

Types

Like their cordwood-burning counterparts, pellet-fired central heaters are available in two main types: furnaces and boilers. As you may recall from chapters 3 and 12, furnaces are designed to be used primarily with a forced hot-air heat-distribution system, while boilers are best matched with a hydronic distribution system.

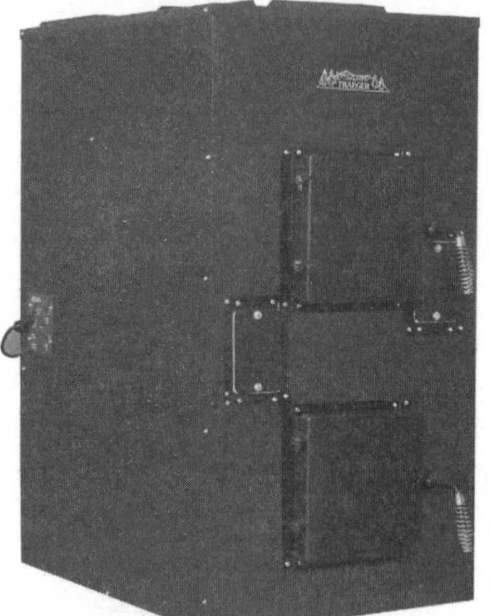

A pellet furnace is highly convenient for a home with central hot-air heating.

Courtesy of Pinnacle Stove Sales, Inc.

Pellet-fired furnaces and boilers look similar to their cordwood-burning counterparts; they're metal boxes, sometimes with a few dials, gauges, or switches on the outside. The presence of a hopper or bin for pellet storage is the only obvious difference. Visual appearance or style is not important in your selection process for these units—it's what's inside the box that counts.

How They Work

Pellet furnaces and boilers function much like pellet stoves, limiting most of the physical labor of operation through automation. An auger feeds pellets to the fire in carefully controlled amounts; the pellets are stored in a bin or hopper, which is usually built into or attached to the outside of the appliance. The auger's speed of rotation determines the amount of fuel added and the amount of heat produced. Not only does this eliminate frequent fueling chores, but also it allows pellet furnaces and boilers to generate just the right amount of steady heat, which is controlled by a thermostat.

Unlike most pellet stoves, virtually all pellet-fired central heaters use a bottom-feed burner design that allows you to use either standard- or premium-grade pellets and sometimes corn. This design causes the pellets to feed into the burn pot from underneath the fire, automatically pushing ash and clinkers out of the pot. The tradeoff is that bottom-feed units are not quite as efficient as top-feed units, and they tend to produce

A pellet boiler is an excellent choice for a home with a hydronic heating system.

Courtesy of Pinnacle Stove Sales, Inc.

Cutaway view of a pellet boiler.

Courtesy of Tarm USA

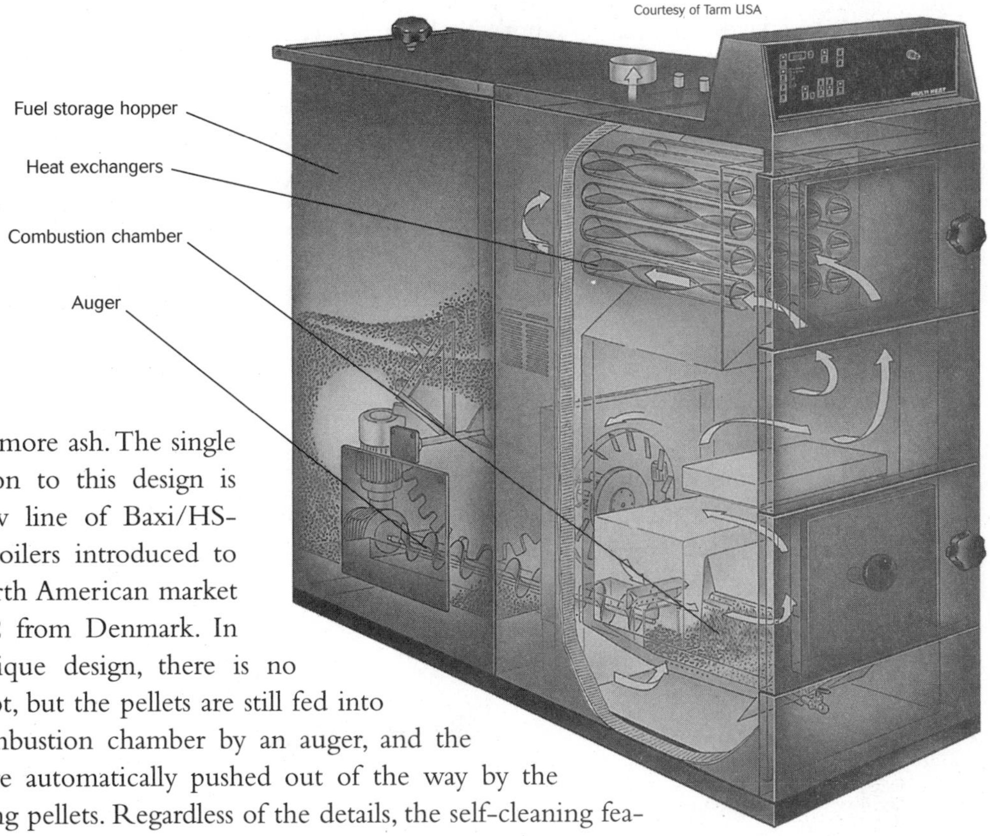

slightly more ash. The single exception to this design is the new line of Baxi/HS-Tarm boilers introduced to the North American market in 2002 from Denmark. In this unique design, there is no burn pot, but the pellets are still fed into the combustion chamber by an auger, and the ashes are automatically pushed out of the way by the incoming pellets. Regardless of the details, the self-cleaning feature on all pellet boilers and furnaces reduces routine maintenance to a minimum. (Low maintenance on a heating appliance located in the basement is a very attractive and important feature.)

Better yet, most pellet furnaces or boilers only need to have their storage hoppers filled every few days (or around once a week on some models), making these appliances an excellent choice for people who want to burn wood (or other biomass) but are away from home much of the time. The frequency of refueling depends on the size of the storage hopper and on the weather. The pellets are combusted in the burn pot (or combustion chamber) with the aid of a blower, which creates an extremely hot and efficient fire with very low emissions. The combustion efficiency of many pellet furnaces and boilers is around 95 percent; their emissions are among the lowest of available solid-fuel-burning heating appliances, and they are an obvious choice in locations where air quality in the winter is an issue.

The blower, which may have up to four speed settings, also directs the combustion gases out of the appliance and into a special 4-inch vent pipe that takes the place of a normal chimney. Before exiting the appliance, the heat from the fire normally passes through a high-efficiency heat exchanger, which warms the air (or water) in your distribution system. The heat exchanger and forced combustion air design allows pellet furnaces and boilers to achieve overall heating efficiencies of around 80 percent.

Sizing a Furnace or Boiler

A pellet furnace or boiler normally burns at high efficiency regardless of the thermostat setting. Because the heat output of these appliances is easily controllable, sizing is not quite as crucial as it is with a cordwood appliance. Some models actually have more than one heat output setting, which allows you to preset heating ranges to match the heating requirements of your home. Nevertheless, most pellet furnace and boiler manufacturers offer several sizes, and it is important to pick the heat output that best matches your home's heating requirements.

Important Features

One key feature to consider when comparing pellet furnaces or boilers is the pellet capacity of the storage hopper. Capacity varies from 80 pounds to 260 pounds, depending on the model. The quantity of pellets you can load into the hopper at one time is a main factor in determining the frequency of refueling. Ideally, you shouldn't have to refuel more than once every few days. Models with extra-large hoppers can sometimes operate for a week without refueling. Another important feature is whether your pellet furnace or boiler has manual or automatic controls. Almost all pellet-fired central heaters have automatic controls and remote thermostats (some units also have automatic ignition systems), important features for an appliance located in the basement. Most of these appliances also offer domestic-hot-water units, outside-air kits, and corn burn pots either as standard features or as options.

The exterior surfaces of most pellet central heaters (with the possible exception of the firebox door) tend to stay relatively cool, reducing the chance of accidental burns from touching them. While the blower fans in some of these appliances may be noisy, this is not as important an issue as with pellet stoves, since furnaces and boilers are located in the basement or a utility room.

Installation

Installing a pellet-fired central heating system is definitely a job for a heating professional—many units weigh about five hundred pounds, and installation requires electrical, plumbing, or sheet-metal expertise. As you plan for installation, be sure to allow for plenty of room to work around your furnace or boiler, and make sure you have an adequate space to store bags of pellets in a handy location. The logistics of moving and handling the pellets for a central heating system are an important part of the picture. If your basement has poor access, you may need to improve it.

As noted previously, pellets offer a substantial advantage over cordwood because they flow easily through pipes and augers. This characteristic offers some wonderful opportunities for automation, but a little ingenuity is called for. Keep in mind that the larger the storage bin, the less frequently you have to refill it. What's more, if you can devise a way to use a chute or pipe to fill a large storage bin in the basement from outside your home, you can eliminate handling individual bags altogether. Of course, this would be an even better arrangement if you could have pellets delivered in bulk by a delivery truck (this service is only available as a demonstration project in parts of New Hampshire and Massachusetts at present). Next, try to design a way for the bulk storage bin in your basement to feed your furnace or boiler automatically with the aid of an auger, blower, or suction system. If you can accomplish all this, you will have a superior, easy-care installation that rivals the convenience of an oil- or gas-fired heating system.

Another factor to keep in mind as you plan is the need for venting. Like other pellet-fired heating appliances, pellet furnaces and boilers need to be vented to the outside, although they do not require a standard masonry or Class A chimney. This allows you flexibility in positioning the special 4-inch type L or PL vent (only the type of vent pipe specified in your unit's installation instructions should be used). On most models, the pipe can be vented directly through an exterior wall or can pass up through the roof or into an existing chimney if one is available (depending on the size of the flue and the condition of the chimney, you may have to reline it). If you run the vent through a wall, be sure to extend the pipe vertically so it clears the roof overhang to eliminate possible staining or damage to your home's exterior finish. The vertical run of pipe will also help your appliance vent naturally in the event of a power failure and minimize the spillage of combustion gases into your living space.

Dual Units

While a pellet furnace or boiler can be installed as a primary heating appliance, another popular strategy is to use them as dual units (add-ons). With this approach, a separate pellet-fired boiler or furnace is installed as an add-on to an existing fossil-fueled central heating system. A dual-unit strategy will result in the most efficient operation for both units, as long as both are properly installed. The add-on strategy is an ideal way to make an existing fossil-fueled system renewable, at least when the pellet-fired unit is operating. Dual systems offer maximum operational and fueling flexibility. If you have an older fossil-fueled boiler or furnace that is nearing the end of its career, taking the dual-unit approach allows you to burn biomass, while adding to the life expectancy of your old oil or gas unit. And, if you have an existing oil-fired unit, you could burn biodiesel in it (see "Biodiesel" on page 42 for details), making your heating system totally renewably fueled.

Operation

In terms of convenience, a pellet-fired furnace or boiler is comparable to oil- and gas-fired central heaters. The primary difference, of course, is the fuel. The operation of most pellet furnaces or boilers is simple. Fill the hopper with pellets, light the fire, adjust the thermostat, and generally that's it. Some units require you to manually light the fire; others are equipped with pushbutton igniters; still others are completely automatic. Be sure to read and follow the operating instructions that come with your unit.

Most pellet furnaces and boilers are equipped with sophisticated electronic controls. Once the fire is burning, the appliance will automatically adjust its heat output to match your heating needs. If it's well below freezing outside, the unit will crank out prodigious heat. If just a little heat is needed, the appliance will reduce its output. On some models, if no heat is required, the unit will slow the burn rate to a very low "pilot" mode until more heat is called for. On other models, the unit will shut itself down and then restart the fire again automatically when more heat is needed.

Most pellet furnaces and boilers have a warning light either on the unit or on the thermostat that will light up when the hopper needs to be refilled. When that happens, pour in a few more bags of pellets and you are all set for a few more days of steady, even heat. An average home that uses a pellet furnace or boiler as its primary heat source will probably require 1 to 1½ bags per day (40 to 60 pounds) of pellets, or around 3 tons per heating season. Larger homes and homes in colder climates may require more.

Maintenance

Routine maintenance for pellet furnaces and boilers is fairly simple (follow the maintenance instructions that come with your unit). The main task is to remove the ashes. This will probably be necessary every few weeks (be sure to store the removed ashes in a metal container with a tightly fitted lid). On some units, the ash pans are large enough that you can burn three or four tons of pellets before you have to empty the pan. This means that you may only have to perform this chore once or twice during the entire heating season. You also may need to clean the heat exchanger after every ton or so of pellets has been burned. In some cases, you can do this with a vacuum cleaner or special brush, while other units are fitted with a handle that allows you to easily clean the heat exchanger without having to open the appliance cabinet. You may want to check the burn pot now and then to be sure it is clean. Most manufacturers advise removing unused pellets from the hopper and auger at the end of the heating season to reduce the chance of rusting. This also helps to eliminate the possibility of having your auger clog up with sawdust fines.

Because they are fairly sophisticated appliances with electronic controls and moving mechanical parts, pellet furnaces and boilers should receive a thorough annual servicing. This generally involves cleaning blowers, changing filters, and checking or lubricating motors and augers. If you enjoy this sort of work, you should be able to follow the annual maintenance instructions that come with your unit and do it yourself. If not, the heating contractor who installed your unit should be able to perform this service work for you. The cost of the service call will probably be about the same as for any standard oil- or gas-burning furnace or boiler. In the case of a major system-component failure, you'll want professional assistance. Most pellet furnaces and boilers do not generate any appreciable amounts of creosote, but the vent or chimney should be checked annually and cleaned if necessary.

Backup

Pellet-fired central heating systems generally use electronic thermostats and other controls, pumps, or blowers. If you lose electric power during a winter storm, you have a problem. If you are lucky enough to have a photovoltaic electrical installation on your home with battery backup, you can wire your central heating system to operate during a power outage for a day or so. But for most people, the best and least expensive approach is to have a backup electrical generator.

Warranties

Pellet furnaces and boilers generally come with limited five-year warranties on the firebox or burner and one- to three-year limited warranties on mechanical and electrical components. As always, be sure to read the warranty on your appliance carefully. Never burn anything other than approved fuels in a pellet furnace or boiler, because doing so will void your warranty.

Where and What to Buy

The best place to buy a pellet-fired furnace or boiler is at a stove specialty shop or a pellet-stove specialty shop. A few heating contractors may offer pellet-fired central heaters. Also, remember that before you buy a central heating unit, it's important to locate reliable, professional service technicians for it in your area. Most pellet-furnace and pellet-boiler manufacturers require that a licensed heating contractor install their appliances.

If you already have an existing hot-air distribution system in your home, then the best match would be a pellet furnace. A pellet boiler would be the obvious complement to an existing hydronic heating system. As with any central heating appliance, long operating life is important, because removing a poor choice is an expensive and complicated procedure. Be sure to

Pellet-Fired Central Heater Pros and Cons

Pros	Cons
Pellet furnaces and boilers can be used in new construction or renovation.	You need a place to store the pellets.
They will heat an entire house, regardless of its interior layout or design.	Some light physical labor is required.
Boilers (and some furnaces) can provide your domestic hot water.	Pellet furnaces and boilers require electricity to operate.
They provide steady, easily controlled heat.	Sitting in front of a pellet furnace or boiler in the basement is not romantic.
They do not require a masonry chimney.	
Most models will burn for several days before you have to refill the storage bin.	
A multifuel or add-on unit offers fuel flexibility and security.	
Pellets are competitively priced and environmentally beneficial.	

look at the manufacturer's history carefully and try to get as much unbiased, third-party feedback as possible. The longer the manufacturer has been in business, the better your chances of getting a quality product. However, since this is a very small market, you won't have many companies to choose among. The primary manufacturers of pellet furnaces or boilers are Traeger (Pinnacle Stove Sales), Harman Stove Company, and BAXI/HS-Tarm.

Price Range

Prices for good central heaters vary, depending on the size and type of unit. A 70,000 Btu hot-air furnace costs around $3,000 plus shipping and installation, while a 130,000 Btu furnace costs around $3,500 plus shipping and installation. A complete hot-air heating system runs between $6,000 and $7,000 installed.

The price of a 150,000 Btu boiler is around $4,500 plus shipping and installation. A complete hydronic heating system costs from $6,500 to $7,500 installed. A high-quality central boiler or furnace is a lifetime investment, so don't compromise on price.

CHAPTER 18

Corn- and Grain-Burning Appliances

People have been burning corn for decades, and sometimes it has had nothing to do with a distracted cook in the kitchen. As I've mentioned before, Midwest farmers used corn as a home heating fuel during the Great Depression. But manufactured heating appliances specifically designed to burn corn didn't come along until the 1980s. Around 1986, the first safety-certified corn-fired home heating appliance was developed by American Energy Systems, Inc. of Hutchinson, Minnesota, according to Mike Haefner, the company's president. "I grew up burning wood, coal, and corn," Haefner says. "Over the years, it's been a natural progression of designing products that burn those fuels. We were looking for fuels that could be locally grown, help protect the environment, and be very economical." Corn met the criteria.

In the first year after the development of his Countryside corn/wood pellet stoves and fireplace inserts, Haefner sold about seven hundred and fifty units, he recalls. Since then, he has revised the design, upgraded and improved the electronic components, and doubled the heat output of the heat exchangers of the units that now kick out around 50,000 Btu, enough to heat an average 2,500-square-foot home. Although the market has had its ups and downs, Haefner says, "Corn burning is unquestionably here to stay."

Corn Power

North America produces about 300 million tons of corn every year, which is theoretically enough to heat over 115 million homes. Heating with a renewable energy crop such as corn can reduce a greenhouse gas (carbon dioxide) by as much as 5.1 tons per home per year. You can achieve reductions like this in your own home by installing a corn-burning heating appliance. Corn-fired appliances now come in almost every size, style, and type. Although there are several different manufacturers, all corn-fired appliances share the attribute that when they are in operation, a subtle scent of baking corn bread drifts through the air. While these appliances are mainly designed to burn corn, a few are also capable of burning wheat, barley, milo, and other grains; agricultural waste, such as cherry and olive pits; and wood pellets. This sort of fuel flexibility makes these units especially attractive in a world of increasingly unstable fossil-fuel supplies and prices. (Throughout this chapter, when I mention corn, the other types of fuels are generally implied as well.)

I also want to mention that the Btu output figure for corn that I cite—7,000 Btu per pound—is somewhat less than the figures used by some people in the industry. I take a conservative approach and follow the guidelines developed at Penn State University. Although this makes corn's heating performance a little less spectacular than is claimed by some, when corn is priced below $2 a bushel, it is still the cheapest fuel available. While corn with 15 percent moisture content is generally considered to be "dry," many corn-burning appliance manufacturers suggest that 11 to 12 percent moisture content is best.

How They Work

Generally speaking, corn-fired appliances operate in the same manner as their pellet-burning relatives. An auger feeds corn, which is stored in a bin or hopper, into the fire in carefully controlled amounts (the corn must be clean, or it may clog the auger). The speed at which the auger turns controls the amount of fuel added and the amount of heat produced. Not only does this eliminate the chore of frequent refueling, but also it allows corn appliances to steadily generate just the right amount of heat, which can be controlled by a thermostat. Most other heating appliances (excepting those that burn pellets) do not have this ability to put out steady heat.

Another great feature is that most corn appliances only need to have their storage hoppers filled once a day (in some cases, only once a week), making these heaters an excellent choice for people who want to burn a renewable fuel but who have to be away from home most of the day. The frequency of refueling depends on the size of the storage hopper and on the weather. The corn is combusted in a burn pot with the aid of a blower, which makes for an extremely hot and efficient fire with very low emissions. The combustion efficiency of many corn units is around 95 percent. Among solid-fuel-burning heating appliances, they have some of the lowest emissions. Corn-fired heating appliances are an obvious choice in locations where air quality in the winter is an issue. The blower, which may have several speed settings, also directs the combustion gases out of the unit and into a small vent pipe that is used instead of a normal chimney. Before exiting the appliance, the heat from the fire normally passes through a high-efficiency heat exchanger, which warms your living space by a variety of methods, depending on the type of appliance. The heat exchanger and forced combustion-air design allows corn appliances to achieve overall heating efficiencies of around 80 percent.

Types

The selection of corn-burning heating appliances runs from freestanding stoves to fireplace inserts and from built-in heaters to furnaces and boilers.

Stoves

The most popular corn stove is the freestanding design. This type of stove, which offers the greatest flexibility in terms of installation, is supported by legs or a pedestal and can be located almost anywhere in the living space of your home. The same floor-protection and stove-location criteria that apply to pellet stoves (specified in chapter 16) are relevant for corn stoves. On average, you can expect a corn-fired stove to burn up to a bushel of shelled corn each day. In an average home, a corn stove will consume 150 to 200 bushels of corn per heating season, depending on the temperature set on the thermostat. That's roughly 1½ acres worth of corn (at 150 bushels per acre).

A freestanding corn stove offers installation flexibility.

Courtesy of American Energy Systems, Inc.

Fireplace Inserts

Like their pellet-fired cousins, corn-burning fireplace inserts can be installed in most functional masonry fireplaces. A metal shield (or shroud) seals off the rest of the opening between the insert and the fireplace mouth. Some corn inserts are approved only for use in masonry fireplaces, while others can also be installed in approved factory-built metal fireplaces. Fuel consumption by these fireplace inserts is about the same as by freestanding stoves.

A corn-burning fireplace insert can be installed in most functional masonry fireplaces.

Courtesy of American Energy Systems, Inc.

Built-In Appliances

A corn-fired built-in appliance offers the warm ambiance of a fireplace without the large expense of a traditional masonry fireplace. Although a noncombustible floor protector is required, these units can be framed in with fairly tight clearances to combustible materials. When faced with brick, tile, or stone, built-ins look generally like a traditional fireplace. Fuel consumption by built-in appliances is about the same as that by freestanding stoves and inserts.

Furnaces and Boilers

A corn-fired furnace or boiler, located in the basement or a utility room, offers none of the ambiance of a hearth appliance, but it does provide steady, dependable heat for a larger home that does not lend itself to heating by a stove. A corn-fired furnace or boiler will heat an average-size home on 200 to 250 bushels of dry shelled corn per heating season. That's about two acres worth of field corn. Some pellet-fired furnaces or boilers can be converted to burn corn.

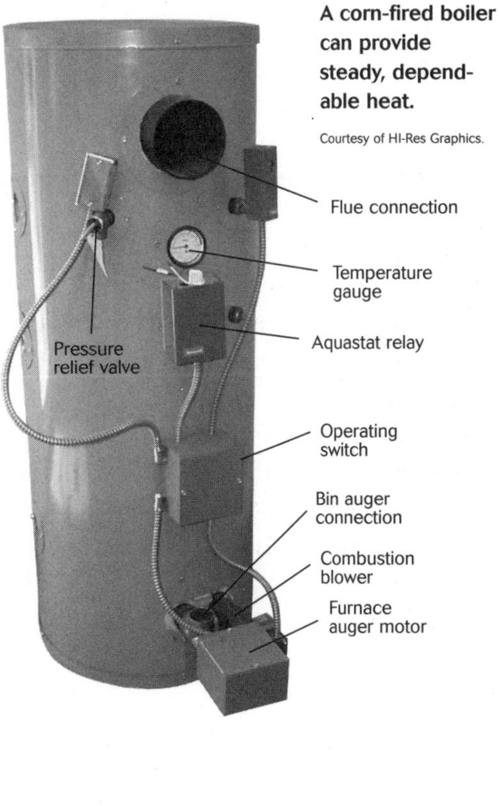

A corn-fired boiler can provide steady, dependable heat.

Courtesy of HI-Res Graphics.

Burner Design

Corn-burning appliances, like their pellet-burning relatives, are further subdivided into top-feed and bottom-feed appliances. Top-feed appliances have an auger that delivers the corn to a tube or chute above the fire pot. In this design, the corn simply falls into the burn pot from above. In a bottom-feed design, the corn is augered to the fire from below or behind the burn pot. Although there are some variations, in the bottom-feed design, the incoming corn pushes the ash and clinkers (fused minerals) away from the burn area. This self-cleaning action of the bottom-feed design reduces the amount of manual cleaning required. While there are some exceptions, top-feed design is more common in corn stoves, while bottom-feed design is more common in corn furnaces and boilers.

Important Features

Features that are important in pellet-burning stoves are also relevant considerations for corn stoves and inserts. These features include storage-bin capacity, sophistication of controls, glass doors, imitation-log sets, blower noise, and ornamental qualities. For corn-fired furnaces and boilers, the main issues are high-quality materials and workmanship, separate thermostats, and storage-bin capacity (the bigger the bin, the less frequent the refueling). Some furnaces and boilers come with large, 14-bushel storage bins that can supply the units for up to ten days.

Sizing the System

Because the heat output of corn-burning appliances is easily controllable, sizing is not quite as crucial as it is with a cordwood stove. Nevertheless, most corn-appliance manufacturers offer several sizes, and it is important to pick the heat output that best matches your home's heating requirements and your weather zone, as well as your expectations and needs.

Installation

As in pellet-burning appliances, the blower-assisted venting feature on corn-fired appliances allows them to be installed almost anywhere, since a standard masonry chimney is not required. However, one of the necessities for the installation of a corn appliance is a nearby electrical outlet. Because of their design, corn appliances require less clearance space than most other types of stoves and can be located as little as one inch from a wall in some cases (follow the clearance guidelines in your stove's installation instructions, as well as all applicable building codes for your location). If you have an existing hearth where you plan to install a corn stove or insert, you probably won't have to worry about floor protection; otherwise, you will need to have a noncombustible floor protector sized according to the installation instructions for your unit.

Corn-fired appliances must be properly vented. A specially manufactured, 3- or 4-inch-diameter type "L" or "PL" vent pipe with a double-wall design is required for most corn burners. Follow the installation instructions about securing and sealing the joints of this vent piping. PL or L vent pipe can be installed through a side wall, through a ceiling and roof, or into an existing chimney. With a through-the-wall approach for the vent piping, most manufacturers recommend at least five feet of vertical piping (or above the eave of the roof) to help combustion gases vent naturally in the event of a power failure.

Operation

Operating most corn-burning heating appliances is generally similar to using their pellet-fired siblings. In many cases, the appliances are identical, with the exception of a special burn pot designed for corn. But there are a few operational differences worth mentioning, particularly the way the fire is started and the frequency of routine maintenance. As with pellet appliances, be sure to keep your fingers away from the auger on a corn appliance to avoid possible injury if it should suddenly start to turn.

Starting the Fire

While details vary somewhat among different corn-burning appliances, the general fire-starting strategy is almost always the same—you don't kindle the fire with corn. That's not a misprint. You don't light the corn, because it's rather difficult to get corn to burn initially. Instead, you place a small handful of wood pellets in the burn pot and then squirt a small amount of starter gel on them. Light the gel and wait until the pellets are burning briskly (this may take a few minutes). Once the fire is well established, you then add corn. The exact details of how this is accomplished can be fairly simple or relatively complex, depending on the appliance. Follow the directions supplied with your particular unit. After the corn is burning, the operation of the appliance is almost identical to a pellet-fired appliance, except for routine maintenance.

Shutting Down

To shut down a corn appliance properly, turn the thermostat or appropriate control knob to the "off" setting and allow the unit to follow its shutdown routine. It may take some time (a few minutes to about an hour) before the fire burns out completely. Once the unit has cooled sufficiently, all motors and blowers will shut down. Just as with a pellet appliance, never turn a corn appliance off by unplugging it. In the event of a power outage, the fire may burn out completely. If so, you will have to relight the fire manually.

There is one final operational hint that can help to mitigate another problem—corn storage. One way of minimizing corn storage over the summer is to try to buy the exact amount that you will need for the heating season, with not a single kernel to spare. In the real world, however, the less corn you have left over at the end of the season, the easier it is to deal with. Any leftover corn should be stored in a tightly sealed, moisture-proof container.

> **Don't Burn Seed Corn**
>
> Warning! *Never* burn seed corn—corn produced for planting in gardens or farm fields—in a corn-fired appliance. Most seed corn is treated with pesticides. Burning pesticide-treated corn may pose a potential hazard to air quality. The even greater potential hazard is the poisoning danger if a young child swallows some of the seed corn.

I have a friend who lives in a residential subdivision and has been happily burning corn as a heating fuel for years. He stores the corn in an open bin in his basement and reports that he has not had any pest problems at all. He does live in a relatively new home with a poured concrete foundation, however. And he has a cat.

Maintenance

Burning shelled corn produces slightly fewer Btu than wood pellets, but it results in more ash. Routine ash removal from most corn appliances is often more frequent than with comparable appliances that burn wood pellets. Probably the single most important factor in minimizing maintenance and maximizing efficient operation is to burn clean, dry corn. Corn with excessive moisture, fines, or pieces of cob will, at the very least, probably clog your feed auger and vent pipe, and it will unquestionably reduce the overall efficiency of the appliance. Burning high-moisture-content corn will actually void many corn-appliance warranties.

Although the routine will vary from one make and model to another, the typical maintenance cycle for corn-burning appliances runs like this. On a daily basis, you will spend a couple of minutes scraping residue from the burn pot, checking the heat exchanger, and monitoring the ash pan to see if it needs to be emptied. Once a week, you may need to shut the unit down and spend about ten minutes cleaning the heat exchangers, emptying the ash pan, vacuuming out the storage hopper, and checking the venting. Once a month, you will probably need to thoroughly clean the appliance and venting. Annual maintenance (performed after the end of the heating season) includes removing all corn from the auger tube and storage bin, thoroughly cleaning the entire appliance and venting, and coating all moving parts with light cooking oil to prevent rusting.

Because they are fairly sophisticated appliances with electronic controls and moving mechanical parts, corn appliances should receive a thorough annual servicing. This generally involves cleaning blowers, changing filters (when applicable), and checking or lubricating motors and augers. You can do this work yourself or hire it out. In the case of a major system-component failure, you'll want professional assistance. Most corn appliances do not generate any appreciable amounts of creosote, but the vent or chimney should be checked annually and cleaned if necessary.

Backup

Because they require electricity to operate, corn-burning appliances need to have some form of backup, especially if they are being used as the primary heat source. Some corn appliances have battery backup packages available as options. And, as I've mentioned before, maintaining an electrical generator is also a good strategy to consider.

Warranties

There is a wide range in warranties for corn-burning appliances, but most generally include limited five-year warranties on major structural

Corn Burner Pros and Cons

Pros

Corn is a homegrown, renewable fuel.

Corn burning appliances can be used in new construction or renovations.

They provide steady, easily controlled heat.

They only need to be refueled about once a day (sometimes less frequently).

They do not require a masonry chimney.

The stove exterior (except the firebox door) remains relatively cool, reducing the risk of accidental burns.

Corn-burning appliances exude the faint smell of corn bread baking in the oven.

Sitting in front of a corn stove or insert is relatively romantic.

Cons

You need a secure place to store the corn.

Most corn-burning appliances require electricity to operate.

The sound of the blower(s) may bother some people.

Some light physical labor is required.

You will need some form of backup if you are going to be away from home for an extended period of time.

Combustion gases may spill into your living space during a power outage.

Sitting in front of a corn-fired furnace or boiler in the basement is not romantic.

components and limited one-year warranties on electrical components. Never burn anything other than approved fuels in a corn stove, because doing so will void your warranty.

Where and What to Buy

As usual, a stove specialty store is the best place to start your search for a corn-burning appliance. Not only do the folks who run these establishments know their products inside and out, but also they can install them in a professional manner.

If you are looking for a space heater, then a corn-burning stove, fireplace insert, or built-in appliance makes the most sense. Which of these you choose depends on the design of your home and your taste and budget. If you need a new or add-on central heater, then a corn-fired furnace or boiler is an obvious match. Regardless of the type of appliance, be sure to look for heavy-duty construction and components. There are about eighteen manufacturers who produce heating appliances that will burn corn, although many of these units are primarily wood-pellet burners with corn-burning options. The principal manufacturers of heating appliances specifically designed as corn burners are American Energy Systems, SnowFlame, and Bixby Energy Systems.

Price Range

The price range of corn stoves and inserts runs from $1,500 to $2,600. Add another $400 to $500 for a typical installation. The price of most corn-fired furnaces is between $3,000 and $3,500, plus installation; corn-fired boilers run around $3,700, plus installation. Generally, corn stoves are somewhat more expensive than cordwood stoves. However, since a corn stove (and a furnace or boiler) can be installed without a traditional masonry or Class A chimney, the *total* installed cost of a corn stove may actually be less than a cordwood stove. Since all corn-burning appliances are complex devices with the potential for mechanical and electrical failures, it makes sense to buy a high-quality unit. Trying to save a few dollars on a "bargain brand" may end up costing you much more in repairs over the long run.

PART FIVE

GEOTHERMAL

CHAPTER 19

Geothermal Home Heating and Cooling

For the rest of this book I'm going to talk about magic. It's miraculous to create a viable and renewable home heating and cooling strategy out of thin air, water, or dirt—and that's what geothermal systems do. This type of geothermal energy relates to the sun rather than hot springs or geysers. (It's interesting that the sun plays a key role in yet another renewable home heating strategy, but you should be used to that recurring theme by now.)

The concept behind geothermal heating is simple: the Earth is a huge heat-storage device. For millions of years, the Earth has been absorbing and storing solar energy in the air, water, and ground. This stored energy offers enormous potential to meet a major portion of our energy needs. The trick up to now, however, has been to figure out practical ways to harvest and use that energy. One of the most successful ways to accomplish that goal is to use a heat pump. Unlike most other home-heating devices, heat pumps are not based on combustion. Instead, heat pumps move heat from one location to another.

Heat pumps have been around since the early 1900s—refrigerators and air conditioners are types of heat pumps. Heat pumps for residential heating were not successfully developed until the 1970s. Although they have been available for over thirty years, heat pumps are still not understood or appreciated by many homeowners, especially those who live in colder climate zones.

Types of Heat Pumps

The term "heat pump" is used to describe a variety of heating systems, and this can be confusing. There are two main types: air-source heat pumps and ground-source heat pumps.

"Conventional **air-source heat pumps** use a refrigerant-to-air heat exchanger. In other words, they use ambient air to heat or cool the refrigerant and ultimately the home," explains Conn Abnee, the executive director of the Geothermal Heat Pump Consortium, Inc. (GHPC). In locations where moderate winter temperatures are the norm, air-source heat pumps work well. In regions where winter temperatures regularly fall below freezing—especially into single digits—a backup heat source (usually electric heat) is necessary, according to Abnee.

"The other main technology, **ground-source**, that we now call GeoExchange, uses a refrigerant-to-water heat exchanger," Abnee says. "This is more efficient because it uses a more constant exchange medium—the Earth—which typically is 50 to 55 degrees [Fahrenheit] just below the surface. This is much more constant than the ambient air." There are three sources of heat for ground-source systems: the ground, well water, and surface water. The ground and well water may not sound like logical places to look for heat, but they actually are an excellent source because their temperature remains relatively constant year-round. The temperature of surface water in rivers, lakes, or ponds varies more than well water, but surface water is a viable heat source in certain situations.

"The main difference between a conventional air-source heat pump and a ground-source GeoExchange system is the fact that you are using a refrigerant-to-air heat exchanger in the first versus a refrigerant-to-water heat exchanger in the latter," Abnee explains. "Conventional air-source heat pumps are a good means of heating and air-conditioning in moderate climates. If they are designed properly and have auxiliary heat, they will work in extremely cold climates as well, but your electric bills will increase dramatically when you have to use the auxiliary heat. GeoExchange technology, on the other hand, can be used worldwide in almost any climate."

Save Oil—Use a Heat Pump

The 650,000 GeoExchange systems presently installed in this country equal 14 million barrels of crude oil saved per year, according to the Geothermal Heat Pump Consortium, Inc. (GHPC). The GHPC is a nonprofit organization (based in Washington, D.C.) that promotes ground-source geothermal technology.

Not surprisingly, GeoExchange systems have attracted quite a lot of interest. The GeoExchange sector has been growing by 30 to 35 percent annually in recent years, according to Abnee. "We estimate there are over 650,000 units installed in this country," he notes. "Our goal is to get over 400,000 units installed annually." By comparison, over 1 million air-source heat pumps have been installed annually nationwide in recent years.

GeoExchange (also called Earth-Energy) systems consume 25 to 50 percent less energy than traditional oil and natural gas systems and up to 70 percent less than electric heat and air conditioning. The GHPC hopes to reduce annual greenhouse gas emissions by 1.5 million metric tons by the year 2005, saving over 300 trillion Btu annually compared to traditional heating and cooling technologies. GeoExchange is the most energy-efficient, environmentally clean, and cost-effective space-conditioning system available, according to the Environmental Protection Agency.

How Heat Pumps Work

You may recall from chapter 1 that heat naturally flows from a warmer area (or substance) to a cooler area (or substance). Heat pumps, however, can force heat to flow in the *opposite* direction, with the aid of a small amount of electricity and a compressor. It's a little like pumping water uphill. A heat pump transfers or "pumps" heat from a natural source such as the air, ground, or well water to the interior of your home during the winter. While we don't think of these sources as being especially "hot," they do, nevertheless, contain useful heat that is continuously replenished by the sun.

Heat pumps can also provide domestic hot water, humidity control, air filtration, and, best of all, cooling during the summer. When the heat-pump process is reversed in the summer, it moves warm air out of your house and into the air or ground (in this case, think of the air or ground as enormous heat sinks). The ability of a heat pump to double as a cooling device is a real advantage, especially in warmer climates, and eliminates the need for a separate air-conditioning system.

One of the most important things to remember about a heat pump is that it operates at such high efficiency (up to 400 percent) because it is primarily *moving* heat from one place to another rather than *creating* heat through some form of combustion. This unique design is what sets all heat pumps apart from most of the other home heating strategies we've looked at previously.

For those who are less inclined to believe in magic, here is the technical explanation of how air-source and ground-source heat pumps operate.

Basic heat pump operation.

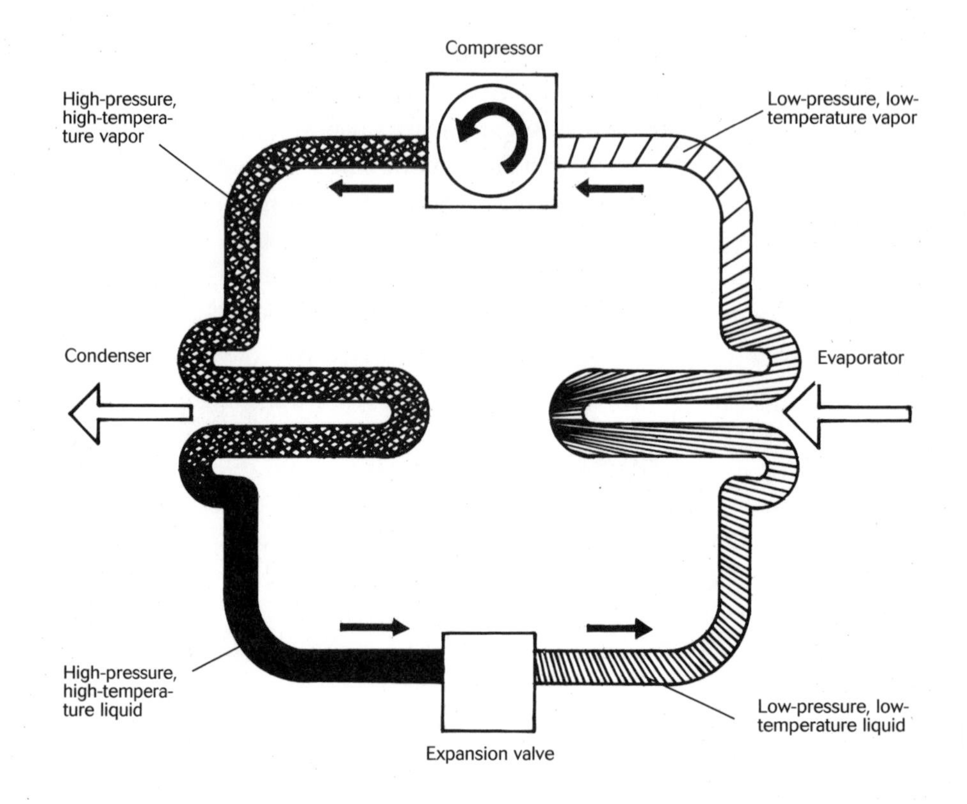

Simply stated, the heat-pump cycle begins as cold liquid refrigerant passes through a heat exchanger and absorbs heat from a relatively low-temperature heat source (air, water, or ground). The refrigerant evaporates as the heat is absorbed, becoming a gas. This refrigerant gas then passes through a compressor, where it is pressurized, raising its temperature to over 160 degrees Fahrenheit. The heated gas then circulates through another heat exchanger, where heat is removed from the gas and transferred to water or air, which is then circulated into your home via a hot-air duct system or a hydronic distribution system (the temperature of the heated air or water is about 100 degrees Fahrenheit). As it loses heat, the refrigerant gas changes back to a liquid. The liquid is cooled as it passes through an expansion device and the heat pump cycle is complete, ready to begin again.

Environmental Considerations

Because heat pumps supply lower-temperature air for distribution than a conventional fossil-fueled furnace, heat pumps tend to run for longer periods of time. For people who are not used to this operational trait, it may appear that the heat pump is "always running." The lower-than-usual temperature of the air that blows out of the heat registers can also be disconcerting to the uninitiated. Yet, a properly designed and installed heat pump will deliver steady heating with less energy consumption than the fossil-fueled competition. This translates into substantial operating-cost savings, especially in areas where fossil fuel prices are relatively high and electricity prices are relatively low. Savings are even greater with variable-speed heat pumps. These units adjust their output to match the actual heating or cooling needs of your home, reducing excessive cycling and wear on the system, especially during mild weather.

A typical air-source heat pump requires just 100 kilowatt-hours (kWh) of electricity to turn the equivalent of 200 kWh of free environmental heat (from the air) into the equivalent of 300 kWh of useful home heat. A ground-source heat pump can be even more efficient. Because heat pumps consume less energy than conventional home heating systems, heat pump use helps to reduce greenhouse gases such as carbon dioxide, sulfur dioxide, and nitrogen oxides. However, the overall environmental impact of heat pumps depends on how the electricity that they consume is generated. (Electricity from renewable sources is obviously preferable to electricity generated by fossil-fueled facilities, especially coal-burning power plants.)

Since there is no combustion involved, heat pumps tend to be very safe, with virtually no risk of fire or combustion gases escaping into your home's living space. Heat pumps are controlled by one or more thermostats. Because they are equipped with a wide range of electrical devices and controls, heat pumps require electricity to function, making them susceptible to power failures. Heat pumps are particularly suited for use in new construction but can also be used in many retrofit situations. Admittedly, heat pumps use a lot of technology that could be largely avoided in a home with an intelligent passive solar design. However, passive solar does not provide air-conditioning, nor does passive solar work well on cold, cloudy winter days in northern climates. Heat pumps, on the other hand, do not suffer from these shortcomings. They are an intriguing marriage between technology and renewable resource.

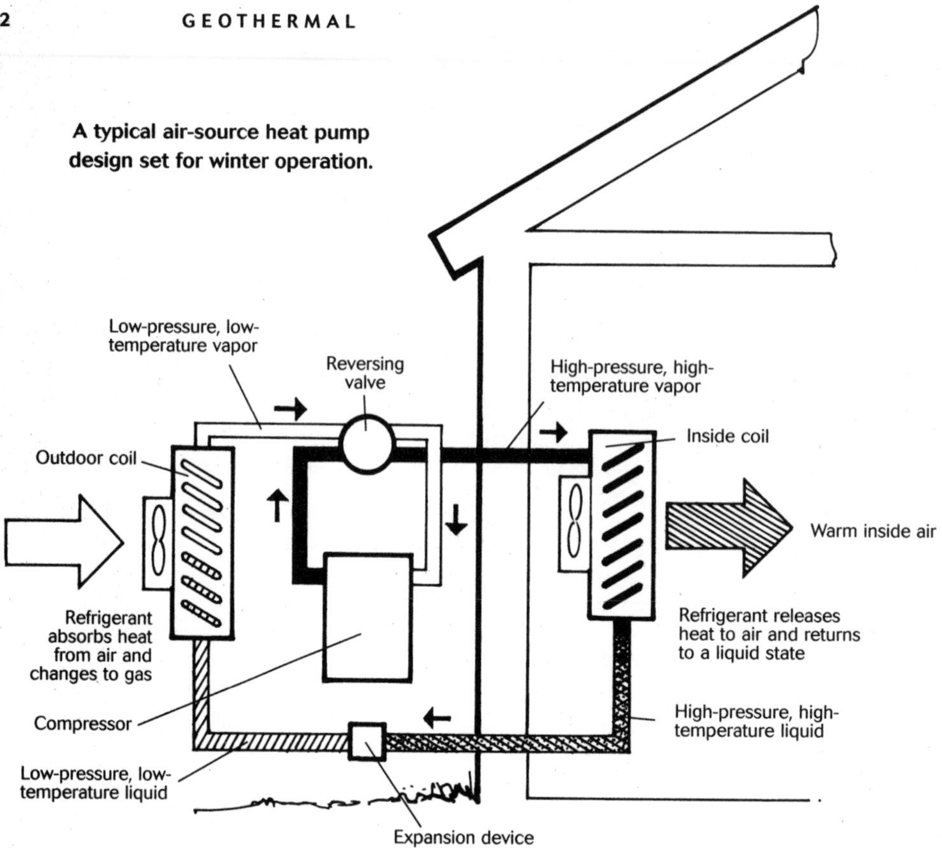

A typical air-source heat pump design set for winter operation.

Heat Pump Systems

On the outside, heat pumps are unexciting; they're rectangular metal boxes with pipes and wires coming out of them. It's what's inside the box that creates the magic. Not all heat pumps are created equal, and there are so many choices. The analogy to a Lego toy set is particularly appropriate when you're combining elements to create an efficient and effective heat-pump system.

Air-Source Systems

The two basic types of air-source heat pumps are one piece and two piece. A one-piece heat pump is self-contained in a single unit. This type of heat pump is often found in commercial settings, such as motel rooms, or is used where a relatively small area in a house or office building is being heated and cooled. One-piece units are often mounted on the roof or in a wall.

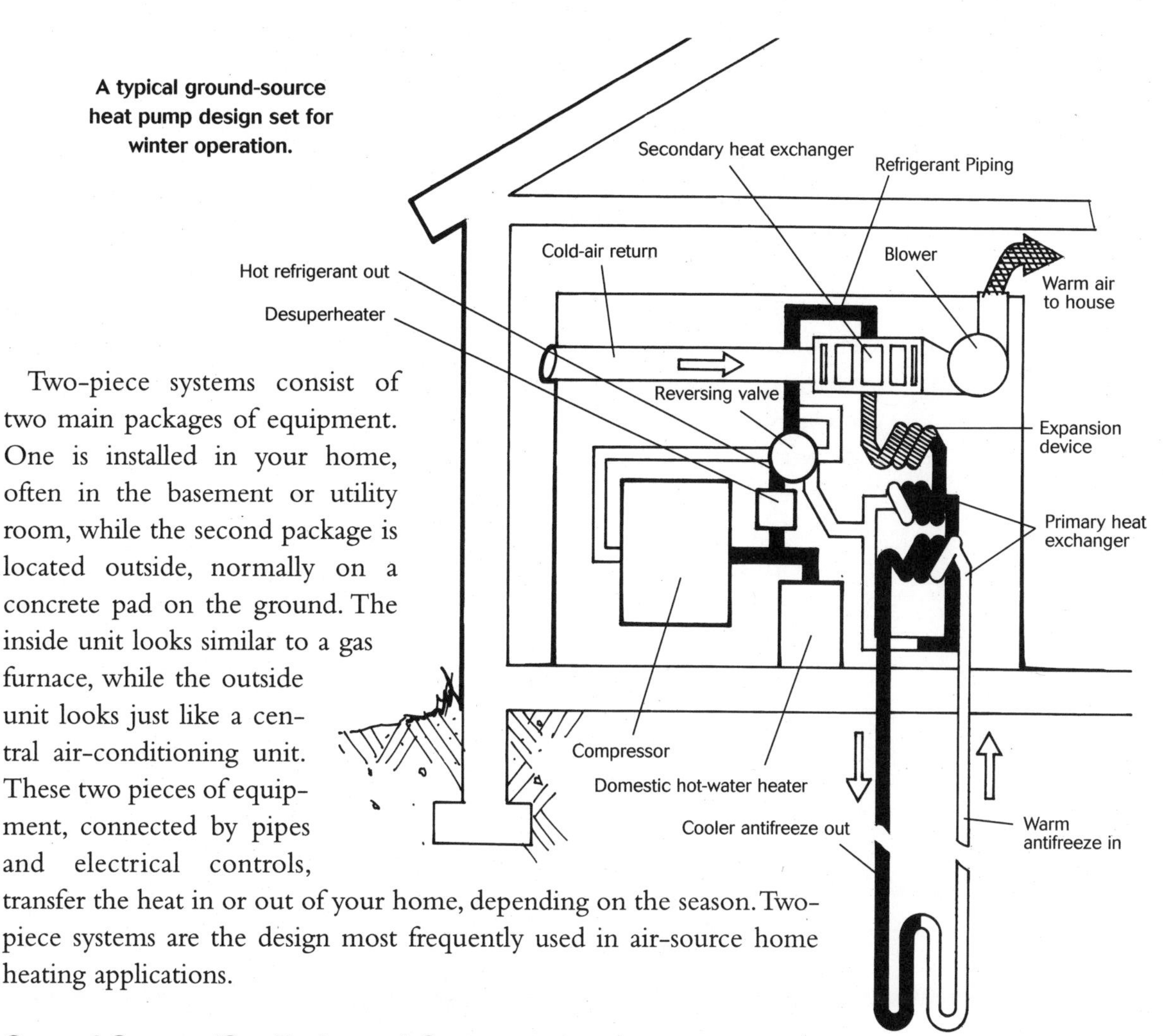

A typical ground-source heat pump design set for winter operation.

Two-piece systems consist of two main packages of equipment. One is installed in your home, often in the basement or utility room, while the second package is located outside, normally on a concrete pad on the ground. The inside unit looks similar to a gas furnace, while the outside unit looks just like a central air-conditioning unit. These two pieces of equipment, connected by pipes and electrical controls, transfer the heat in or out of your home, depending on the season. Two-piece systems are the design most frequently used in air-source home heating applications.

Ground-Source (GeoExchange) Systems

Unlike an air-source heat pump, where one heat exchanger (and frequently the compressor) is located outside, a ground-source heat pump unit is located entirely inside your home, often in the basement or utility room. This is especially important in northern climates, where extremely cold winter temperatures and snow can have an adverse impact on heat-pump equipment located outdoors. Part of the heating system is located outside, however, and that is the underground piping that contains the liquid heat-exchange medium. The type and configuration of the outside piping system is the main factor that differentiates various ground-source heat pump designs.

Heat Pump Components

Regardless of the type of heat source, all heat pumps have certain components in common. You don't need to understand all the technical detail of how these components operate, but it's a good idea to know what they are and what they do.

The Compressor

The heart of all heat pumps is the **compressor.** The compressor circulates the refrigerant, which is the lifeblood of the system. Simply put, the compressor squeezes the molecules of the refrigerant together, raising the temperature of the refrigerant. Heat pumps may use either a reciprocating compressor or a rotary compressor.

Reciprocating

A reciprocating compressor is a durable, time-honored veteran that works well with a wide range of refrigerants. This is the most common type of compressor that you'll hear chugging along in older heat pump and refrigeration systems.

Rotary

The rotary compressor is a more recent development. It's worthwhile to seek out a heat pump with a rotary compressor. This device offers smoother operation, higher efficiency, and lower power consumption than a reciprocating compressor. Plus, a rotary compressor has fewer moving parts, is more compact, and operates more quietly. In particular, the lower noise level makes a rotary compressor a better choice for many heat pump installations, since compressor noise can be a real problem with heat pumps. Another advantage of a rotary compressor is that it does not require an accumulator (a part that prevents liquid refrigerant from entering and damaging a reciprocating compressor) or a crankcase heater (reciprocating compressors need these to prevent lubricating system damage).

A variation on the rotary compressor is the scroll compressor, which offers even higher efficiencies at lower temperatures as well as extremely quiet operation, compact size, and solid reliability. Scroll compressors also eliminate the need for accumulators and crankcase heaters.

Expansion Device

The **expansion device** is the system component that releases the pressure created by the compressor. This causes the temperature of the refrigerant to drop, and it again becomes a low-temperature vapor/liquid mixture.

Valves

All heat pumps have valves, but the most interesting is the reversing valve, sometimes referred to as a **four-way** (or magic) **valve**. The magic valve controls the direction of flow within the heat pump system, depending on whether heating or cooling is needed. In the heating mode, refrigerant flows one way; in cooling mode, it moves in the opposite direction.

Heat Exchangers

All heat pumps have heat exchangers of one type or another. These devices (sometimes referred to as coils) are made up of loops of tubing, often with fins to increase surface area. The coils and associated fans that circulate air through the coils are what actually heat and cool your home. In most designs, heat exchangers are also the component that transfers heat from the air or ground to the refrigerant. In a "split system," the heat exchanger for the living space is separate from the heat pump and is added to an existing hot-air furnace.

Supplementary Heat

In the case of most air-source (and some ground-source) heat pumps, supplementary electric heating elements are located in the main air-circulation space (also called a plenum) for times when the outside air temperature falls too low for the heat pump to meet the total heating requirement of your home. In the case of a "split system," the existing hot-air furnace becomes the supplementary heater during extremely cold weather.

Thermostats

Most residential heat pump systems use a "two-stage heat/one-stage cool" indoor thermostat. Stage one calls for heat if the temperature falls below the preset level. Stage two calls for heat from the supplementary heat component if the indoor temperature continues to fall below the desired level.

A variety of outdoor thermostats that respond quickly to outdoor temperature conditions are also available. Using an outdoor thermostat can result in more efficient heat pump operation. Programmable thermostats are also available from most heat pump manufacturers, and can result in operating savings from temperature setbacks at specific times during the day or night.

Capacity Rating

The most common heat-pump sizing measurement is the **ton,** which has nothing to do with how heavy the pump is. This archaic term is a holdover from the days when refrigeration units were used mainly to produce ice for old-fashioned iceboxes that kept food cool. A "one-ton" unit could make a ton of ice in a day. In today's terms, a one-ton heat pump can generate 12,000 Btu of cooling per hour at an outdoor temperature of 95 degrees Fahrenheit, or 12,000 Btu heat output at 47 degrees Fahrenheit.

The **coefficient of performance (COP)** is a measure of a heat pump's efficiency. This is determined by dividing the heat pump's energy output by the electrical energy needed to run it. A heat pump's cooling efficiency is indicated by the **energy-efficiency ratio (EER)**, which is the ratio of the heat removed (in Btu per hour) to the electricity required (in watts) to operate the unit.

The **heating-season performance factor (HSPF)** is a measure of the total heat output (in Btu) of a heat pump over the entire season divided by the seasonal power consumption (in watts). Overall cooling efficiency is indicated by the **seasonal energy-efficiency ratio (SEER)**, which is a measurement of the cooling efficiency of the heat pump over the entire cooling season. SEER is determined by dividing the total cooling provided (in Btu) by the total energy used (in watt-hours). The main point to remember about these ratings is that the higher the rating, the better the heat pump.

The Distribution System

Most heat pumps move the heat that they generate into the living space in your home through a warm-air ductwork distribution system (certain types of hydronic systems can also be used). It's important that the ductwork be carefully designed to ensure comfortable temperatures throughout the heating and cooling seasons. But equally important, airflow needs to be sufficient to prevent damage to the heat pump. Inadequate airflow may eventually lead to compressor failure—and a large repair bill. Too much airflow, on the other hand, can lead to a noisy and drafty system. A standard ductwork system for a central hot-air heating system may not be large enough for a heat pump. The finer points of proper ductwork design are complicated and require special knowledge and training. This is why it is so important to have an experienced heat pump contractor design the system and perform the installation.

Heat Pump Contractors

The installation of a heat pump is definitely a job for a professional, and finding a knowledgeable, experienced, and competent installer is extremely important. In some areas, this can be a challenge, because the support infrastructure for heat pumps is not evenly distributed nationwide. At the very least, one quality to seek out is enthusiasm for heat pump technology. Avoid a heating contractor who advertises heat pump installations but tries to talk you into some other type of system. Another pitfall to avoid is the "lowest-bidder syndrome." Obviously, you don't want to pay too much for a heat pump, but paying too little may lead to difficulties later on. Since the majority of heat pumps manufactured by reputable companies offer similar performance and quality, most problems with operation can be traced to improper or poor-quality installation. A bid from a reputable installer will cover the realistic cost of proper installation. A contractor who is trying to drum up business by offering unrealistically low estimates may cut corners on the installation. Compare several bids, and always ask for references. Find

A Reliable Heat Pump Contractor

Your prospective contractor should be able to:

- Calculate the heating and cooling load for your house and be able to explain it clearly to you
- Ensure that your electrical system is capable of accommodating the increased load of the heat pump
- Provide full details on the operation and warranties for your unit
- Provide a service contract on the entire system
- Guarantee the installation work

Heat Pump Pros and Cons

Pros	Cons
Heat pumps generally have lower operating expenses than other conventional heating systems.	They require more careful maintenance than some other heating systems.
They can both heat your home in the winter and cool and dehumidify it in the summer.	They require electricity to operate, a disadvantage in locations where power outages are frequent.
They offer steady, even heat.	They do not quickly heat or cool a home.
Since they operate without combustion, heat pumps present very little fire hazard, and there are no combustion gases to spill into your living space.	Sitting in front of a heat pump is not romantic.
They take up less space than a typical oil-fired furnace and oil storage tank installation.	
They do not need a chimney.	
They are very clean.	

out how many systems the contractor has installed. If previous customers are experiencing system problems or failures, beware.

Another important, but often overlooked, aspect of choosing a contractor is service. If you have a problem with your heat pump in January, you'll want quick emergency service. Does your prospective installer have a twenty-four-hour emergency phone number, and does someone always answer when you call the number? When you are checking references, ask specifically about service and how promptly it was performed. Also, be sure your contractor is certified by the manufacturer to service the particular brand of heat pump you are having installed. Another useful qualification to look for is certification by the **International Ground Source Heat Pump Association** (www.igshpa.okstate.edu) or the Refrigeration Service Engineers Society (RSES).

Which Heat Pump Is Best?

This is not an easy question to answer. It's difficult to identify one best system when there are so many possible choices. Even people working in the heat-pump industry express conflicting opinions about which heat pump offers the most advantages. Air-source heat pump manufacturers point to high repair costs associated with ground-source system failures. GeoExchange manufacturers cite air-source system limitations in extreme climates and higher operating costs for the backup electric heat needed for many air-source packages.

I'll offer this rule of thumb. If you live in a moderate climate where winter temperatures below freezing are infrequent, an air-source heat pump is a better choice. If you are located in a region where subfreezing winter temperatures are the norm, then choose a ground-source (GeoExchange) heat pump system. Note that site-specific circumstances in some situations (such as lack of space for an in-ground piping system or lack of usable ground water) will suggest a different strategy. Also, be aware that certain models of heat pumps of both major types are designed for optimum performance in warmer climate zones, while other models are designed to work better in colder zones. Discuss all of your options with your heat pump installer; it's the best way to ensure that the system you choose will meet your needs.

CHAPTER 20

Air-Source Heat Pumps

Air-source heat pumps deliver on the promise that you can heat and cool your home out of thin air. It sounds outlandish, but it's true. When they were first introduced to the market in the 1970s, air-source heat pumps were installed mainly in the South. Designs have improved since then, and air-source heat pumps now are the most popular type of residential heat pump in the United States. Reflecting that popularity, about 1.5 million units were sold in 2001, and millions of air-source heat pumps are in operation nationwide. You are likely to see a lot more heat pumps in the future, since approximately 26 percent of all new single-family homes in the United States are now being equipped with heat pumps.

These systems work extremely well in moderate climates to renewably heat and cool your home, but because they rely on outside air as their source of heat, these systems don't work as well in extreme climates. In regions where extended periods of subfreezing temperatures are common, air-source heat pumps are not the best approach, although they do perform better than other electric-heating systems in those regions.

An outside unit of a heat pump always contains the outdoor heat exchanger and usually the compressor.

Courtesy of Bryant Heating & Cooling Systems

Air-source heat pumps have three main components: the outside unit, which always contains the outdoor heat exchanger (or coil); an inside unit, which contains the indoor heat exchanger; and the heat distribution system. The compressor is usually located in the outside unit, but in some instances is installed indoors. Air-source heat pumps offer a number of advantages. First, the heat source—the air—is abundant and free. Second, air-source heat pumps are the least expensive type of heat pump, and initial installation costs can be quite reasonable. Last, but not least, air-source heat pumps have been on the market for a long time. This means that manufacturers have had a chance to work out all the bugs and glitches years ago, making air-source heat pumps dependable and long lasting.

Of course, there are also some disadvantages. Operating costs for air-source heat pumps tend to be higher than for ground-source heat pumps. Frost build-up can also be a problem (more on this issue later). In order for an air-source heat pump to work efficiently, the difference between the desired temperature in your home and the temperature of the heat source should be as small as possible. This is why air-source heat pumps are not very cost effective in extremely cold climates where the temperature differential can sometimes be enormous.

Types of Air-Source Heat Pumps

Air-source heat pump systems can be air-to-air systems or air-to-water systems (these terms relate to the type of heat-distribution system inside your home). The most common is the air-to-air system. In this design, heat is extracted from the air and then moved inside or outside your home, depending on the season. This type of system also can remove dust, pollen, and other allergens from the air in your home.

An air-to-water design uses a hydronic distribution system inside your house to distribute heat or remove heat. (Don't confuse this with ground-source heat pumps that use water as an *outside* heat source). With an air-to-water system, the heat pump removes heat from the outside air and then transfers it to the water in your home's hydronic distribution system during the winter. In the summer, the process is reversed: the heat pump extracts heat from the hydronic system inside your home and then moves it outside, where the heat is dissipated into the air. Air-to-water air-source heat pumps are relatively rare.

Air-source heat pumps are further subdivided into add-on or all-electric categories. Add-on heat pumps are designed for use in conjunction with a source of supplementary heat, such as oil, gas, or electric furnaces—creating a dual-fuel system. All-electric heat pumps, on the other hand, are equipped with their own electric-resistance supplementary heating system. This electric-resistance system is essentially an electric furnace and also is sometimes referred to as backup or emergency heat. While this electric heat can be used when your heat pump fails, it is useless if your electricity goes out during a winter storm.

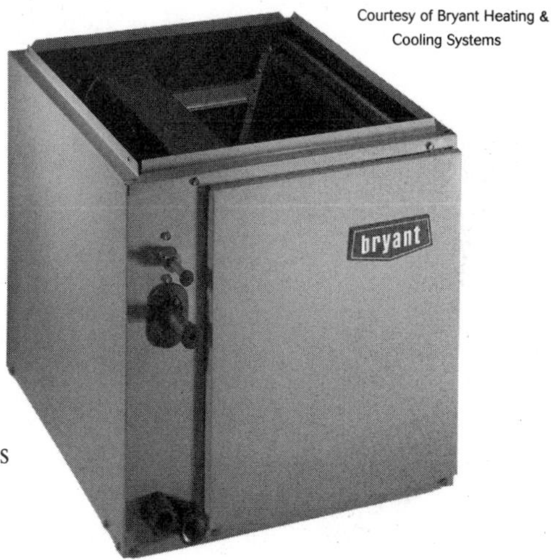

An inside unit of a heat pump contains the indoor heat exchanger.

Courtesy of Bryant Heating & Cooling Systems

Choosing the Correct Heat Pump

Unlike ornate cast-iron stoves, heat pumps are not designed to be pieces of living room furniture. External appearance has nothing to do with selection criteria. It's performance that counts. You can easily judge a heat pump's performance by checking the efficiency ratings. The heating efficiency for air-source heat pumps is indicated by the heating-season performance factor (HSPF), which is the ratio of the seasonal heating output (in Btu) divided by the seasonal power consumption (in watts). Cooling efficiency is indicated by the seasonal energy-efficiency ratio (SEER), which is the ratio of the seasonal heat removed (in Btu per hour) to the seasonal power consumption (in watts). The simplest approach is to look for an Energy Star label, which is awarded to units with an SEER of 12 or greater and an HSPF of 7 or greater.

If you live in a warm climate, you should give more weight to the cooling efficiency rating. If you live in a predominantly cool climate, the heating efficiency is more important. It's also helpful to know that some models of heat pumps are specifically designed to operate in colder climates, while others are optimized for warmer regions.

Another important selection criterion for air-source heat pumps is noise. The operating sounds from the outdoor unit of some heat pumps are annoyingly loud. This is a problem if the unit is located under your (or your neighbor's) bedroom window. All heat pumps have sound ratings, which are expressed in units of **bels**. Select a heat pump with an outdoor sound rating of 7.6 bels or less if possible. The lower the rating, the quieter the unit. Since this is a logarithmic scale, even slight differences in the rating can make big differences in sound level. For example, the sound of a large diesel truck 10 feet away from you would have a rating of 9.2 bels, while the noise level of a refrigerator (with its compressor operating) is rated at 7.0 bels.

Sizing Your Heat Pump

Sizing an air-source heat pump to match the heating requirements of your home is extremely important. Installing an oversize unit will result in excess energy consumption, inferior humidity control, and shorter equipment life due to short cycling. What's more, there's no advantage in paying the higher cost of installing a larger system if the size isn't needed. An undersized heat pump, on the other hand, will cause your supplemental heat to operate too frequently, increasing your utility bills.

The same criteria for sizing most heating appliances also apply to air-source heat pumps, including the dimensions of your home, size and number of windows and doors, insulation, energy efficiency, and local weather. If you are planning to use your heat pump for cooling, you will also need **cooling load calculations** to determine how much cooling capacity your heat pump will need for your home.

The most popular methods for determining heating and cooling loads are based on methods and data developed by the American Society of Heating, Refrigerating and Air-Conditioning Engineers (ASHRAE). Your heat pump contractor should be familiar with these methods and be able to explain the process to you.

Another important sizing factor is called the balance point. As outside temperature drops during the winter, so does the capacity of an air-source heat pump to absorb outdoor heat. Unfortunately, this occurs just at the time when you need more heat in your home in order to stay comfortable. This can lead to problems if the outdoor temperature drops too far. At the **balance point**, the heat pump's capacity is equal to the heat loss of your house. Typical balance-point temperatures fall between 27 and 35 degrees Fahrenheit. When the outdoor temperature falls below the balance point, the heat pump cannot supply all the heat needed, and supplementary heating is required. In colder climates where heating is the main goal, the correct size for your system is a compromise between heating and cooling, with emphasis on heating. Trying to hit the desired balance point with the right combination of heating and cooling capacity is tricky and is best left to a professional heat pump system designer.

Installation

Air-source heat pumps are particularly well suited to new construction, because you can plan in advance for the various system elements. Replacement or add-on systems are also possible in most situations. When you are planning an air-source heat pump installation, keep in mind that the two units (or three, if the compressor is located indoors) of an air-source heat pump are connected by electrical wiring and refrigerant lines, offering some flexibility in unit placement. On the flip side, however, refrigerant lines should be as short and straight as possible.

Locating the Outdoor Unit

The outdoor unit should be placed on a stand that is anchored to a concrete pad (unless the manufacturer specifies a different strategy). The stand should raise the unit from 1 to 2 feet above ground level to minimize loss of efficiency caused by snow accumulation. (The idea is to keep the unit above the snow.) The outdoor unit should be placed where it is protected from prevailing winter winds (which can intensify frosting problems). At the same time, the airflow around the unit should not be too restricted. Low shrubbery can provide wind protection. Don't plant the bushes too close, however, or they may eventually block airflow when they grow larger. Maintain a horizontal distance of at least 30 inches between the outside unit and any obstruction.

Also, the outdoor unit should not be located under the drip line of the roof, in order to prevent water, ice, and snow from falling on and possibly damaging the fan and compressor. An ideal location for an outdoor unit would be on the south side of your house, shaded by deciduous trees in the summer. During the winter, the sunlight passing through the bare branches would help warm the unit. It's a good idea to place refrigerant lines to the outside unit in an insulated conduit to minimize heat loss and prevent condensation. The distance between the outdoor and indoor unit should not exceed about sixty feet. Locate the unit away from windows or other areas where its noise might annoy your neighbors. Some units make noise due to vibration, which can be minimized by mounting the unit on a noise-absorbing base.

Locating the Indoor Unit

A basement, laundry room, closet, crawl space, or attic are all potential locations for the indoor unit of a heat pump, as long as they are not too far from the outdoor unit and there is sufficient space for working on the equipment. Nevertheless, the best location is one that is as central in your home as possible, allowing for more efficient air circulation in the hot-air distribution system.

Distribution System

Although it's possible to use either a hot-air distribution system or a hydronic distribution system with an air-source heat pump, most designers and installers prefer a hot-air system. If your heat pump is going to be an add-on or conversion unit, the existing hot-air ductwork should be carefully inspected and evaluated to ensure that it is adequate. For proper heat pump operation, airflow should be from 400 to 500 cubic feet per minute per ton of cooling capacity. In many cases, part or all of older ductwork for a fossil-fueled furnace will not be large enough and will need to be modified. If your home does not already have a hot-air distribution system, installing new ductwork can be a challenge, but it's not impossible. The difficulty or ease of installation depends on the design and construction details of your home. Ask your heat pump installer for system-design, equipment, and installation estimates and then decide for yourself if the project makes financial sense.

Electrical Service

Virtually all air-source heat pumps will require a 200-amp electrical service. While most new homes have 200-amp or larger electrical service, many older homes do not. Upgrading the service in homes with 60- or 100-amp services can be an expensive project, and upgrading the main electrical service entrance and circuit breaker panel is a job for a professional electrician. If you need to upgrade your service, include the cost of this work in the total estimate for your heat pump installation.

Operation

Air-source heat pumps have three operating cycles: the heating cycle, the cooling cycle, and the defrost cycle.

The Heating Cycle

The **heating cycle** for air-source heat pumps begins when cold liquid refrigerant passes through the outdoor heat exchanger (or coil) and absorbs heat from the air. The refrigerant evaporates into a gas as the heat is absorbed. The gas then passes through the reversing ("magic") valve and into the compressor, where the refrigerant is pressurized, raising its temperature. The heated gas then passes through the reversing valve again before it circulates through the indoor heat exchanger, where heat is removed and distributed throughout your home. When it loses heat, the refrigerant changes back to a liquid. The liquid is cooled as it passes through an expansion device and begins the process again.

Cutaway view of an outside unit of an air-source heat pump.

Courtesy of Bryant Heating & Cooling Systems

The Cooling Cycle

In the **cooling cycle**, the refrigerant travels through the cycle in the opposite direction from that of the heating cycle. During the cooling cycle, most air-source heat pumps also dehumidify the indoor air. The moisture in the air passing over the indoor heat exchanger condenses on the coil's surface and collects in a pan below the coil. The collection pan normally empties into a house drain.

The Defrost Cycle

If the outdoor temperature falls near or below freezing while an air-source heat pump is operating in heating mode, moisture in the air will condense and freeze on the outside coil. This frost buildup reduces the coil's ability to transfer heat. Eventually, the frost must be removed. To accomplish this, the heat pump will automatically switch to defrost mode. In the **defrost cycle**, the reversing valve switches the heat pump into cooling mode, which sends hot refrigerant to the outdoor coil to melt the frost. At the same time, the outdoor fan is turned off, assisting the defrosting process. Unfortunately, while this is happening, the heat pump is sending cool air into your home, and it is sometimes necessary for the system to turn on the supplementary heating until the defrost cycle is completed.

Defrost Control

Defrost controls fall into two categories: time-temperature and demand. Time-temperature controls turn the defrost cycle on at specific predetermined intervals, such as every thirty or sixty minutes, depending on the climate and the system's design. Unfortunately, this can result in unnecessary defrost cycles, which can be hard on your system. Demand controls, on the other hand, only initiate the defrost cycle when frost is actually detected. I recommend that you ask for demand defrost controls on your air-source heat pump.

Other Operating Considerations

The thermostat on a heat pump should be set at your desired temperature and not be readjusted. If you use a programmable setback thermostat, the night setting should not be lower than 65 degrees Fahrenheit. This is because a heat pump takes much longer to raise house temperature to daytime comfort levels than a conventional fossil-fueled or wood-heated appliance. Heat pumps are more efficient at maintaining comfortable temperatures but have to work very hard to bring up temperatures from a "cold start." If your heat pump has a high-efficiency variable-speed fan motor (highly recommended), operate the system on the "auto" fan setting on the thermostat. Finally, a heat pump tends to run for longer periods than a conventional fossil-fueled system because its heat output is lower. It takes a while for some people to get used to this characteristic.

Maintenance

Studies have shown that the expected service life of air-source heat pumps tends to be shorter in colder northern climates than in milder southern regions because climate affects the total hours of operation. In colder regions, heat pumps spend more time heating than cooling, and the heating cycle puts more stress on the heat pump. However, careful installation and maintenance can have as much or more impact on the service life of the system. In general, servicing the many components of a heat pump should be left to a specially trained technician. An annual service call to inspect and possibly adjust your unit is strongly recommended. The best time to service your heat pump is at the end of the cooling season but before the heating season begins. Since heat pumps are more sophisticated than most standard heating systems, service costs may be slightly higher.

There are some basic maintenance tasks that you can—and probably should—perform yourself. It's hard to overemphasize the importance of careful maintenance of a heat pump. Dirty filters, coils, and fans reduce airflow through the system. If neglected for too long, this can damage the compressor. Check the filters monthly and clean or replace them as required by the manufacturer's guidelines. Clean the coils with a vacuum or brush at least once a year (always turn the unit off first). Check your ductwork annually to be sure it is free of excessive dust buildup or any other obstructions and also to ensure that carpets or furniture are not blocking the vents and registers. Keep the outdoor unit clear of snow, ice, and debris, and keep shrubs pruned back at least 18 inches (30 inches is better) from all sides of the outdoor unit. Be sure to read the maintenance instructions that come with your unit and keep them handy to the appliance for future reference.

Backup

Because they require electricity to operate, air-source heat pumps need to have some form of backup, especially if they are the primary heat source. As I've mentioned before, having an electrical generator for backup is a good strategy to consider.

Warranties

Heat pumps installed in the 1970s and 1980s generally have had life expectancies of about fifteen years. Units installed more recently can be expected to last for about twenty years. Most air-source heat pumps are covered by a one-year warranty on parts and labor, with an additional limited five-year warranty on the compressor (usually for parts only). Extended warranty options are sometimes available. The terms of warranties vary from one manufacturer to another. One way to assure a high-quality installation is to ask for a performance guarantee on the entire installed system, rather than just a limited warranty on the heat-pump equipment itself. Many installers offer service contracts as well.

Where and What to Buy

You'll find a wide variety of air-source heat pumps available through most heating, ventilation, and air-conditioning (HVAC) contractors. Ask two or three contractors to estimate the installation cost for the air-source heat pump you are interested in, so you'll be able to make a comparison.

Your main selection criterion for an air-source heat pump should be performance. Be sure to look for the Energy Star label, which is awarded to units with an SEER of 12 or greater and an HSPF of 7 or greater. The bright yellow EnergyGuide label should also display this information. The Consortium for Energy Efficiency (CEE), a nonprofit organization that promotes energy-efficient products, has established recommended efficiency levels for heat pumps. Also, look for equipment that is certified by the Air-Conditioning and Refrigeration Institute (ARI), a nonprofit organization that rates the performance of heat-pump equipment. Certified equipment carries the ARI seal.

There are too many manufacturers of air-source heat pumps to list them all here. Some big-name companies are familiar to most people as air-conditioner manufacturers, while other companies only make heat pumps. Any manufacturer that produces units that have been awarded high Energy Star ratings is worthy of your consideration. Consult the Energy Star Web site at www.energystar.gov for listings of qualifying manufacturers.

The ARI seal indicates that a heat pump has been certified by the Air-Conditioning and Refrigeration Institute.

Courtesy of the Air-Conditioning and Refrigeration Institute.

Price Range

In most areas, air-source heat pumps are competitive with conventional heating and cooling systems in terms of price. Air-source heat pumps cost about $500 per ton capacity plus installation but can be more expensive. If you need to install a hot-air distribution system or modify an old one, your installation costs will be higher. Ditto if you need to upgrade your electrical service. Don't forget to ask if rebates are available from your utility for energy-efficient heating and cooling equipment.

Since operating costs for a heat pump are normally lower than those for conventional heating appliances, you should be able to save on energy costs. The relative savings, however, will depend on whether you are currently using oil, propane, natural gas, or electricity to heat your home and on their price in your area. When you run a heat pump, you will use less fossil fuel

Air-Source Heat Pump Pros and Cons

Pros

- The heat source, air, is free and abundant.
- They are the least expensive type of heat pump to install.
- They have been on the market for many years and are dependable and reasonably long-lasting.

Cons

- The operating costs are higher than for other types of heat pumps.
- In moist, cold temperatures, the outside unit will frost, requiring a defrosting cycle.
- Outdoor units can be noisy and need to be sited carefully.
- During extremely cold weather, air-source heat pumps need to rely on a back-up heat source.
- They need electricity to operate, a disadvantage in locations where power outages are frequent.

but more electricity (except when compared with an electrically heated house, in which case you'll definitely use less electricity with a heat pump). The payback period for your investment in an air-source heat pump may be anywhere from a few years to more than a decade.

So does an air-source heat pump make sense for you? If you live in a moderate climate zone, there's a good chance it might. On the other hand, if you live in an older home with a small (under 200 amp) electrical service entrance and have an oil- or gas-fired hydronic baseboard heating system, the cost of installing a new air-source heat pump system, including new hot air ductwork, may be prohibitive. Crunch the numbers and then decide. If you do opt to have an air-source system installed in your home, you can relax in the knowledge that while fossil fuel prices will fluctuate wildly in the future, your fuel source will always be available—and free.

CHAPTER 21

Ground-Source Heat Pumps

In the last chapter, I showed you how to create renewable heat for your home out of thin air. In this chapter, I'm going to tell you how to heat your home with dirt and water. Of course, I'm talking about ground-source heat pumps. It's ironic that we've literally been walking around on an unlimited source of heat for thousands of years without fully appreciating its potential. We're finally starting to catch on. Ground-source (GeoExchange) heat pumps are not as common as their air-source cousins, but they have many advantages that a growing number of people are beginning to appreciate. One of the major remaining obstacles to wider use of these systems is that many people simply aren't aware of them.

Savings from the Ground Up

The Earth is a huge energy-storage device that absorbs 47 percent of the solar energy that strikes its surface. This represents over five hundred times more energy than we need every year. Ground-source heat pump systems take some of this heat from the ground during the heating season at an efficiency of around 400 percent, and then return it during the cooling season. Spectacular efficiency figures like this can result in substantial savings. According to the U.S. Environmental Protection Agency (EPA), ground-source systems can save you 30 to 70 percent in heating costs and 20 to 50 percent in cooling costs, compared to conventional fossil-fueled systems. What's more, studies have shown that approximately 70 percent of the energy used in ground-source systems is renewable energy from the Earth. Ground-source systems work well in almost any climate, especially more extreme climates where air-source heat pumps are less cost-effective.

How do these systems perform their magic? A ground-source heat pump uses the Earth, ground water, or surface water as sources of heat in

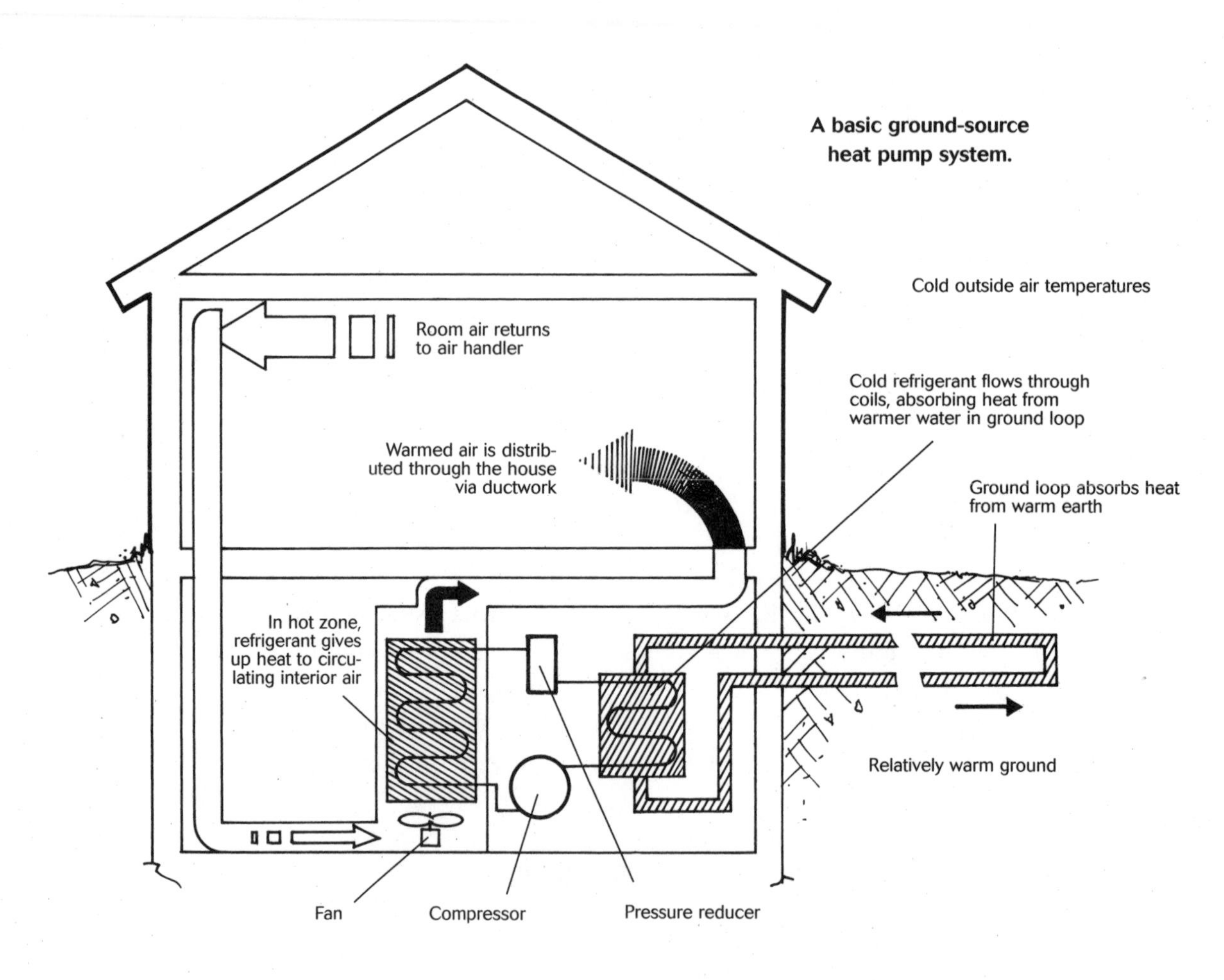

A basic ground-source heat pump system.

the winter and as a "sink" for heat removed from your home during the summer. Heat is extracted from the Earth by a liquid, such as ground water or antifreeze. The heat pump components amplify this heat and transfer it to the air in your home. The process is reversed for cooling. Like their air-source counterparts, ground-source heat pumps can also filter your air and provide humidity control as well.

Ground-source heat pumps have three main components: a system of underground piping outside the house; a heat pump unit inside the house; and a heat distribution system, also located in the house. Since the heat pump unit in a ground-source system is located only inside a home, there are no problems caused by frost buildup or damage from ice, snow, or other severe weather conditions.

Some ground-source heat pump units contain all of the usual elements—

blower, compressor, heat exchanger, and condenser coil—in a single cabinet. But in a split-system ground-source installation, the coil is added to an existing hot-air furnace and uses that system's existing blower. In this case, the heat pump is the primary heater, while the existing furnace can provide supplementary heat during extremely cold conditions.

Types of Ground-Source Heat Pumps

Ground-source heat pumps can be open system or closed loop. These terms refer to the design of the piping system located outside the home. An **open system** uses the latent heat in a body of water (usually a well but sometimes a pond or stream) as its heat source. The water is pumped from the well or other source to the primary heat exchanger in the heat pump, where heat is extracted. The water is then discharged into a pond or stream or into a separate (or sometimes the same) well. In a **closed loop,** however, heat from the ground is collected by means of a continuous loop of piping buried underground. An environmentally safe antifreeze solution circulating through the piping absorbs heat from the surrounding soil. The antifreeze solution is then drawn into the primary heat exchanger in the heat pump, where heat is extracted. A variation on the closed loop is a **direct-expansion loop**, also called a **DX system**. In a DX system, refrigerant runs directly from the heat pump to the underground piping without passing through a heat exchanger, increasing operating efficiency by 10 to 15 percent.

Operation

The operation of a ground-source heat pump is similar to that of an air-source heat pump except for the source of heat and number of operating cycles. Unlike air-source heat pumps, ground-source systems only have two operating cycles: the heating cycle and the cooling cycle (there is no defrost cycle).

The Heating Cycle

In the heating cycle, the ground water or antifreeze mixture absorbs heat from the Earth while circulating through an underground piping system. The water or antifreeze is then brought into the heat pump inside your home, where it passes through a refrigerant-filled heat exchanger. The heat is transferred to the refrigerant, which evaporates into a gas as the heat is absorbed. The gas then passes through the reversing ("magic") valve and into the compressor, which pressurizes the refrigerant, raising its temperature. The heated gas then passes through the reversing valve again before it circulates through the condenser coil, where heat is removed and distributed throughout your home. After releasing its heat, the refrigerant then passes through the expansion device, cooling even further before it returns to the primary heat exchanger to begin the cycle again.

The Cooling Cycle

The cooling cycle is the reverse of the heating cycle and moves heat from inside your home to be dissipated into the ground from the water, antifreeze mixture, or (in a DX system) the refrigerant.

Choosing the Correct Heat Pump

As with air-source heat pumps, exterior appearance of ground-source systems is not important. It's performance that counts. The heating efficiency of ground-source heat pumps is indicated by their coefficient of performance (COP), which is the ratio of heat provided (in Btu) per unit of energy input (also in Btu). Their cooling efficiency is indicated by the energy-efficiency ratio (EER), which is the ratio of the heat removed (in Btu per hour) to the electricity required (in watts) to operate the unit. Look for the Energy Star label, which indicates a heating COP of 2.8 or greater and a cooling EER of 13 or greater.

The same general selection guidelines that apply to air-source heat pumps apply to ground-source units, except for noise of the outside unit, since there isn't any (ground-source heat pumps are so quiet that many homeowners don't even realize they are running). In addition, the same developments in compressors, motors, and electronic controls for air-source units have resulted in improved efficiency in ground-source units as well; for highest efficiency operation and lowest operating costs, look for models with two-speed scroll compressors and variable-speed indoor-fan motors.

Sizing Your Heat Pump

The same criteria for sizing most heating appliances also apply to ground-source heat pumps, including the dimensions of your home, size and num-

Hot Water on the Side

In some ground-source systems, a special heat exchanger (sometimes called a **desuperheater**) is used to heat domestic hot water. On some models, the heating of domestic hot water takes place only under certain operating conditions, while on others it occurs on demand. Excess heat is always available during the cooling mode and often available in the heating mode when the heat pump is not working at full capacity. The ability of a ground-source heat pump to provide domestic hot water as an "extra," in addition to space heating and cooling, makes these systems especially attractive.

While it's possible to meet all of your heating needs with a ground-source heat pump in many locations, the most cost-effective sizing strategy in colder climates is to install a system that can provide 80 to 90 percent of your home's total maximum space-heating and domestic-hot-water load. Occasional peak heating loads that exceed the system's design can be met by a supplementary heating system. High-efficiency ground-source heat pumps with two-speed compressors can meet most average loads on low speed and only need to switch to high speed under severe winter conditions. This system flexibility leads to higher operating efficiency and lower operating costs. Your heat pump contractor can design an appropriately sized system to meet your needs in your particular climate zone.

ber of windows and doors, insulation, energy efficiency, and local weather. If you are planning to use your ground-source heat pump for cooling, you will also need cooling-load calculations. Your heat pump contractor should be able to perform all of these calculations for you.

Correctly sizing your ground-source heat pump to match the heating requirements of your home is very important. Oversize units will be inefficient; undersized units will require too much supplemental heat. The output of a ground-source system does not vary significantly during the winter because its heat source (earth or ground water) maintains a fairly constant temperature year-round. This allows a ground-source heat pump to provide most of the heat required for your home, often with enough extra capacity to provide domestic hot water as well. A well-insulated, 2,000-square-foot home will probably require around a 3-ton heat pump. The actual size of your system should be within 15 percent of the calculated load.

Installation

Unlike air-source heat pumps, which simply use outside air, ground-source systems require a well or a loop system of piping to collect and dissipate heat. Consequently, ground-source heat pumps are more complicated and more expensive to install than their air-source counterparts. Ground-source heat pumps are well suited to new construction because you can plan for the various system elements in advance. This is especially true for the underground piping system outside your home. On a construction site, the additional excavation work for the piping will not be particularly intrusive. In a renovation, on the other hand, installing the necessary underground piping for a ground-source system can be quite a challenge. The contractor will have to avoid disturbing existing water, sewer, and underground utility lines. Excavating is also difficult if you have a carefully landscaped yard with extensive flower beds, mature trees, and other plantings.

Open Systems

An open system is the simplest to install.

In locations where groundwater is plentiful, an open-system heat pump design may be your most efficient option. An open system is the simplest to install and uses water from a well (or sometimes from a pond, lake, or river) as its heat source. After the heat has been extracted from the water by the heat pump (or added to it during the cooling cycle), the water is discharged into a drain tile, stream, river, pond, or lake. This is sometimes referred to as the **open-discharge method.** Another open-system strategy is to put the "used" water from the heat pump into a **discharge well.** In some designs this may be a separate well. In others, the discharge well is the same well from which the water was first drawn; this is called a **standing-column well.** A standing column well is very deep. The water for the heat pump is drawn from the bottom of the well while the return water is discharged into the top of the well.

Discharge pipe

Well water

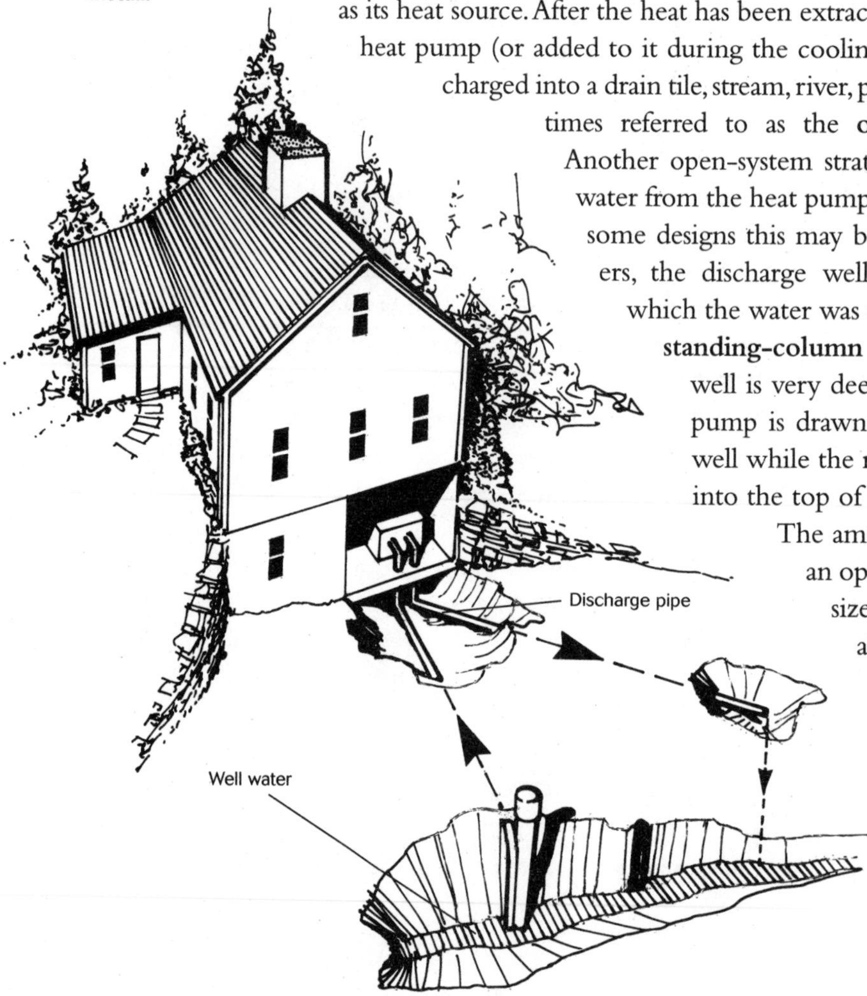

The amount of water required in an open system depends on the size of heat pump, but generally ranges between five and twelve gallons per minute when the heat pump is operating. It's extremely important that the well is capable of producing more than this amount of water on a steady basis, especially if you use the same well for domestic water. Regardless of the method of disposal of the "used" water, no pollutants are added. The only change is a slight decrease or rise in temperature, depending on whether the heat pump is operating in the heating or cooling cycle.

Poor water quality can cause serious problems in open systems. Don't use water that contains excessive particles of organic matter, which can quickly clog a heat pump. Have the water tested for acidity, hardness, and iron content before using it in your system. Your heat pump contractor or the equipment manufacturer can tell you what standard of water quality is acceptable. Your local zoning or environmental regulations may restrict or prohibit open systems. Check with local officials early in the planning stage.

A vertical-loop installation is a good choice where yard space is limited.

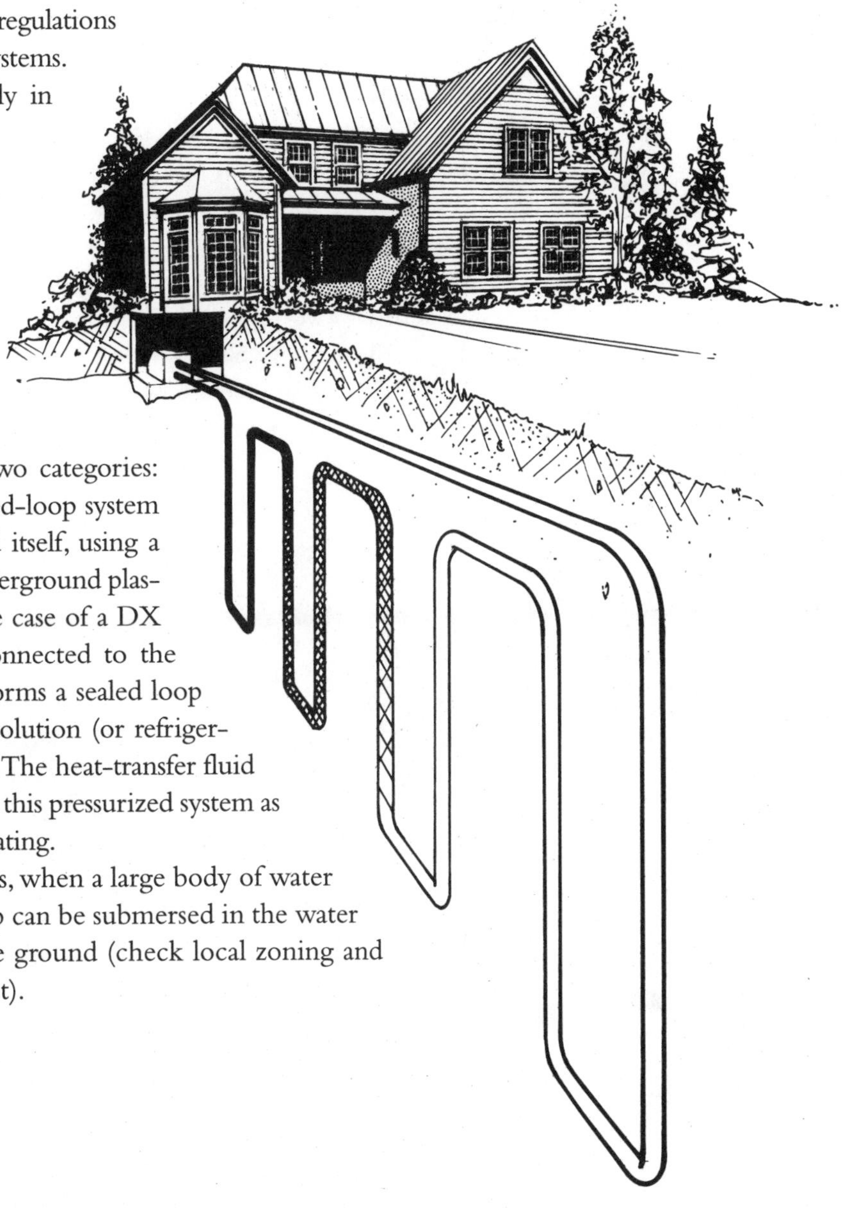

Closed-Loop Systems

In locations where ground water is not available in sufficient quantities, a closed-loop design may be a better strategy as a heat source for your heat pump. Closed-loop systems are subdivided into two categories: vertical and horizontal. A closed-loop system extracts heat from the ground itself, using a continuous loop of special underground plastic pipe (or copper pipe in the case of a DX system). When the pipe is connected to the heat pump in your home, it forms a sealed loop through which an antifreeze solution (or refrigerant in a DX system) circulates. The heat-transfer fluid constantly recirculates through this pressurized system as long as the heat pump is operating.

Under certain circumstances, when a large body of water is located nearby, a closed loop can be submersed in the water instead of being buried in the ground (check local zoning and environmental regulations first).

Vertical Loop

In locations where yard space is at a premium, such as in a suburban subdivision, a vertical loop may be the best design. U-shaped loops of special piping are inserted into 6-inch-diameter holes that have been bored 60 to 200 feet deep. The length of the loop piping depends on your home's heating and cooling load, soil conditions, climate, and landscaping. For every ton of heat-pump capacity, from 270 to 350 feet of piping is required. The vertical holes may need to be backfilled with a special grouting material to ensure good heat transfer. Your heat pump contractor will advise you on the best methods for your area. Vertical loops are generally more expensive to install than horizontal loops.

Horizontal Loop

In rural locations where yard space is not an issue, the horizontal-loop design is more popular. In this strategy, the special plastic pipe is placed in trenches that are 3 feet to 6 feet deep. (The depth depends on the climate zone as well as the number of pipes in each trench.) For every ton of heat pump capacity, 400 to 600 feet of pipe is required. The most common design is to lay two pipes side by side in a trench, but more pipes can be installed if the size of your lot is limited. An alternative piping design that maximizes limited land area uses a spiral- or coil-shaped pipe.

Regardless of the pipe design, the pipe must be made from a special,

A horizontal-loop installation is the most popular and cost-effective strategy where yard space is not an issue.

tough polyethylene plastic with fused (rather than glued or clamped) joints to ensure that they won't leak. Properly installed piping that meets these specifications should last from twenty-five years to more than fifty years. When this piping is backfilled, the backfill material must be carefully tamped to ensure good heat transfer. After installation, the trenches need to be restored by planting grass seed or sod. Vertical loops, on the other hand, result in minimal damage to your lawn or other landscaping.

Locating the Heat Pump

A basement, crawl space, or utility room are the most obvious locations for the heat pump itself, but you can install it elsewhere as long as there is sufficient space for you or a service technician to work on the equipment. The best location is one that is as central in your home as possible; this will allow more efficient air circulation in the hot-air distribution system.

Distribution System

Although it's possible to use either a hot-air distribution system or a hydronic distribution system with a ground-source heat pump, most designers and installers prefer a hot-air system. If your heat pump is going to be an add-on or conversion unit, the existing distribution system should be carefully inspected and evaluated to determine if it will work properly with the heat pump. In many cases, part or all of an older distribution system for a fossil-fueled furnace or boiler will not be adequate and will need to be modified or replaced. If your home does not already have a hot-air distribution system, installing new ductwork can be a challenge. The difficulty or ease of installation depends largely on the design and construction details of your home. Get a complete installation estimate from your heat pump installer and then decide for yourself if the project makes financial sense.

Electrical Service

As with air-source heat pumps, ground-source systems will require a 200-amp electrical service. While most new homes have 200-amp or higher electrical service, many older homes do not. If you need to have your electrical service entrance and circuit breaker panel upgraded, be sure to include the cost of this work in the total estimate for your heat pump installation.

Maintenance

Ground-source heat pumps require slightly less maintenance than air-source systems, since the heat pump itself is located inside your home rather than outdoors. Nevertheless, as with any heat pump, routine preventive maintenance is very important. A dirty filter, coil, or fan will reduce airflow through the system, and over time, reduced airflow can damage the compressor. Check the filter monthly and clean or replace it as required by the manufacturer's guidelines. Clean the coil with a vacuum or brush once a year (always turn the unit off first). Check your ductwork annually to be sure that it is free of excessive dust buildup or any other obstructions and that the vents and registers are not blocked by carpets or furniture.

In an open system, mineral deposits can build up inside the heat exchanger. A service technician can clear away the deposits with the use of a mild acid solution. A closed-loop system is sealed from the possible buildup of mineral deposits from ground water. An annual service call to inspect and possibly adjust your ground-source unit is strongly recommended. The best time to service your heat pump is at the end of the cooling season but before the heating season begins. Since heat pumps are more sophisticated than most standard heating systems, service costs may be higher.

Backup

Because they require electricity to operate, ground-source heat pumps need to have some form of backup, especially if they are being used as the primary heat source. If you have an electrical generator, you will not need any other backup system.

Warranties

Ground-source heat pumps have a useful life span of twenty to twenty-five years. They last longer than air-source units because the compressor is protected from the severe fluctuations of the outdoor environment and is consequently subjected to less stress. Most ground-source heat pumps are covered by a one-year warranty on parts and labor, with an additional limited five-year warranty on the compressor (usually for parts only). Extended warranty options are sometimes available. The terms of warranties vary from one manufacturer to another. One way to assure a high-quality installation is to ask for a performance guarantee on the entire installed system in addition to the limited warranty on the heat pump equipment. Many installers offer service contracts as well.

Where and What to Buy

Some (but not all) heating, ventilation, and air-conditioning (HVAC) contractors offer ground-source heat pumps. Because ground-source systems require careful design and installation of the underground piping as well as the heat pump itself, contractors who specialize in these systems are your best choice for information and installation work. Seek estimates from two or three contractors (if you can find that many in your area) for installation costs for your system. If possible, try to find a contractor who is certified by

Ground-Source Heat Pump Pros and Cons

Pros

- The ground is readily available as a source of heat in most locations.
- Operating costs for ground-source heat pumps tend to be low.
- They have the lowest life-cycle cost of any conventional heating system.
- They can provide some or all of your domestic hot water.
- They are very quiet.
- They have a very low impact on the environment.

Cons

- They are more expensive to purchase and install than most other types of heating systems.
- Ground water in sufficient quantities may not be available in some locations.
- Ground water may have too much mineral or iron content to be usable.
- Locating and fixing an underground leak can be difficult and expensive.
- Replacing an underground piping system at the end of its useful life is also expensive and disruptive.
- Because of the excavating work involved, ground-source systems are not as well suited for renovations as for new construction.
- Some new products in ground-source systems may not be as dependable as time-tested air-source components.
- They need electricity to operate.

the International Ground Source Heat Pump Association (IGSHPA). Check their Web site at www.igshpa.okstate.edu. Some heat pump installers are not as familiar with ground-source systems as they are with air-source systems, so you need to choose a contractor carefully. Most of the complaints that I have heard about ground-source heat pumps have been related to poor system design or installation.

Your main selection criterion for a ground-source heat pump should be performance. Look for equipment that is certified by the Air-Conditioning and Refrigeration Institute (ARI). Certified equipment carries the ARI seal. The Energy Star Web site is another good source of rating information for ground-source heat pumps (go to www.energystar.gov/products/forhome.shtml and follow the "Heating & Cooling" link to "Geothermal Heat Pumps"). Be sure to look for the Energy Star label on the heat pump you are investigating. The following manufacturers have been awarded high Energy Star ratings for their ground-source heat pumps: American Geothermal DX, ClimateMaster, ECONAR Energy Systems, ECR Technologies, FHP Manufacturing, GeoTech Systems, Global Energy & Environmental Research, HydroDelta, The Trane Company, and WaterFurnace International.

Price Range

The price of a ground-source heat pump system can vary enormously due to a wide range of site-specific factors. Open or ground-water systems tend to be less expensive than closed-loop systems. The IGSHPA estimates an average system will run about $2,500 per ton of unit capacity. This estimate does not include the possible costs of electrical work, ductwork, water hookup, or other modifications to your home that might be necessary. The total installed cost of a complete ground-source system can be up to twice the cost of a conventional fossil-fueled system with add-on air-conditioning. As a general rule of thumb, the cost of one of these systems can run from $7,500 to more than $15,000. (If you have to drill two deep wells, it could be a lot more.) At these prices, it doesn't hurt to ask whether rebates are available from your local utility for energy-efficient heating and cooling equipment. Your utility may also be able to supply a list of local installers, dealers, and contractors.

On the flip side, a recent EPA study showed that ground-source heat pumps have the lowest life-cycle cost of all conventional home heating systems available today. So, if you are able to get past the initial installation costs, you can end up saving money in the long run. The study also showed that ground-source systems have the lowest impact on the environment when compared to conventional systems.

Solar-Assisted Heat Pumps

Solar-assisted heat pumps are a relatively new combination of renewable home heating technologies that offer some interesting possibilities. The main idea with this strategy is to use solar power to assist the heat-pump process. This can be accomplished by using an active liquid solar heating system during the winter to preheat the ground water before it passes into the heat pump. Another approach is to use photovoltaic panels to generate some of the electricity needed for the various elements of the heat pump. With these strategies, you should be able to reduce your operating expenses and improve operating efficiency.

While this all sounds great in theory, there are some problems in practice. The most obvious is that the sun doesn't always shine in many locations, especially during the cold, cloudy winter months when you need it the most for heating purposes. The other main disadvantage is cost. Ground-source heat pumps are expensive enough to begin with in terms of equipment and installation costs. Active solar heating systems are not cheap either, and the combination of the two could be prohibitively expensive. It would be prudent to carefully analyze the long-term savings in operating expenses versus installation costs to see if this strategy makes financial sense.

Notes from the Field

I've talked to a lot of people about ground-source heat pumps. Along the way I've picked up some interesting feedback. This information is anecdotal, not definitive, but may still help you in making decisions about ground-source heat pumps.

In colder climates such as northern New England, the preferred heat source is definitely a deep well. Closed-loop systems have proved disappointing for some homeowners. The consensus seems to be that the problems were due to poor system design or improper installation (or both) by contractors who didn't know what they were doing. Also, some of the materials being used in closed-loop systems are relatively new and have not yet withstood the test of time.

Many of the people I've spoken with report that they experienced some initial problems with their ground-source heat pump system. But after the installer worked out the bugs, the homeowners have been very happy with their system. One of the biggest complaints (in northern New England) is a lack of qualified installers and service technicians. One Vermonter I spoke with said that he didn't think about that issue when he had his system installed. He noted that, in retrospect, he might not have chosen a ground-source heat pump if he had understood how hard it was going to be to get prompt and affordable service. He also noted that, in his opinion, a ground-source heat pump was probably not a good choice for people who never pay attention to their heating system—until it breaks down. I agree. A heat pump, like most electromechanical systems, does require routine preventive maintenance to keep it in top operating condition. If you totally ignore your heat pump, you're asking for trouble.

Is there a ground-source heat pump in your future? If you are planning to build a new home, now is definitely the time to look into the possibilities of a ground-source system, while your options are still open. If you are contemplating an add-on or replacement system, you need to look at a wide range of factors such as your home's structure and design, type of existing heating system, available yard space, and supply of underground or surface water. Assuming that the project looks feasible from a technical standpoint, you will also want to do some calculations to see if it will be cost-effective. If the numbers add up, you may wish to proceed. Even if the numbers don't add up, you might decide to go ahead anyway because you want to be environmentally responsible. If so, you have my respect and admiration. And, you can have the added fun of telling all your friends and relatives that you are heating your home with dirt from your backyard.

Epilogue

This country has an enormous problem with unsustainable energy policies, and unsustainable home heating practices are a significant part of that problem. But from reading this book, you have learned that there are many viable alternatives—so many, in fact, that you may be suffering from a case of alternatives overload. Should you build a passive solar home, or retrofit your existing home with active or passive solar elements? Should you heat your home with wood in a stove, fireplace, furnace, boiler, or masonry heater? Perhaps your heating appliance should burn pellets, corn, or biodiesel? Should that appliance be attached to a hot-air or hydronic distribution system? Or does a heat pump make more sense?

The answer to these questions, as I suggested at the beginning of the book, will be different for each person's situation. You find the answers by weighing all the advantages and disadvantages of the various approaches: open the Lego box, play with the parts, and have fun developing the unique solution that matches your lifestyle and needs.

What unique solution did I end up with? When I started writing this book at the end of 2001, I was trying to decide what kind of heating system would best replace the aging oil-fired boiler in the basement of our home in Weybridge. What to do with the fireplace in the living room was also on my mind every time I built a fire that ended up setting off the smoke alarms. As time went by, I realized that we have several options for using renewable heating appliances in our home. We've considered installing either a pellet insert or a woodstove in the fireplace. We've also looked into the possibility of replacing the old boiler in the basement with a new pellet-fired unit. Our south-facing garage roof would be a perfect location for an active solar domestic-hot-water collector. But because all of these strategies involve a considerable investment, so far, they haven't happened.

The one strategy that didn't involve a large outlay of cash up front was burning biodiesel fuel in our old boiler. So we tried it. After two heating seasons without incident, I can say that this experiment was an unqualified success. Of all the strategies I've covered in this book, I must admit that I was the most impressed with biodiesel. Using biodiesel is so simple. Although it is being promoted mainly as a vehicle fuel, I think biodiesel has equal or even greater potential as a renewable home heating fuel. There's no appliance conversion needed as long as you have an oil-fired boiler or furnace. You can burn as little or as much of the stuff as you want (or can afford), as long as you start out with small quantities and gradually increase the percentage as the sludge (if any) in your oil tank and fuel lines dissolves. In our case, the shift to a biodiesel blend was totally transparent, meaning we simply couldn't tell the difference. And there's no fire to stoke, no wood to handle, no ashes to clean out, no limitation on how long we can be away from our home in the winter. It doesn't get much easier or more convenient than that.

The main obstacles to wider use of biodiesel as a heating fuel are price, supply—and politics. The three are related. Biodiesel is more expensive than number 2 heating oil and is simply not available from most fuel dealers. And because it's not subsidized by the Federal government the way oil is, biodiesel isn't able to compete on a level playing field. Of course, the same problems apply to virtually all renewable energy strategies in this country—the playing field is tilted decidedly in favor of the oil, coal, natural gas, and nuclear industries, making it extremely difficult for renewables to compete.

But it doesn't have to be this way. I urge you to get involved in promoting renewables at every opportunity on the local, state, and national level. Everybody has a role to play. Become involved. Push for legislation that encourages renewable energy. If you don't succeed the first time, try again. Turn up the heat on your elected representatives and keep their feet to the fire. While you're at it, be sure to keep your own feet warm with a renewably fueled home heating system—whatever it may be. And enjoy the satisfaction and pleasure of helping the environment while staying comfortably warm in the winter and knowing that you are part of the solution rather than part of the problem.

ORGANIZATIONS AND ON-LINE RESOURCES

General Information on Sustainable Heating Technologies

Energy Star®,
U.S. Environmental Protection Agency (EPA)
401 M Street, S.W. (6202J)
Washington, DC 20460
Phone: (888) 782-7937
E-mail: info@energystar.gov
Web site: www.energystar.gov
An excellent, up-to-date listing of the most efficient appliances, lightbulbs, windows, home and office electronics, and much more.

Home Power magazine
P.O. Box 520
Ashland, OR 97520
Phone: (800) 707-6585
E-mail: hp@homepower.com
Web site: www.homepower.com
An informative magazine that is a wonderful resource for information on renewable energy in general. It occasionally publishes articles on solar heating systems for residential applications.

National Renewable Energy Laboratory (NREL)
1617 Cole Boulevard
Golden, CO 80401
Phone: (303) 275-3000
E-mail: webmaster@nrel.gov
Web site: www.nrel.gov
The leading center in the United States for renewable energy research.

Rocky Mountain Institute
1739 Snowmass Creek Road
Snowmass, CO 81654-9199
Phone: (970) 927-3851
E-mail: outreach@rmi.org
Web site: www.rmi.org
A nonprofit organization that fosters the efficient and restorative use of resources to create a more sustainable world. Energy efficiency is one of this organization's many areas of expertise.

U.S. Consumer Product Safety Commission (CPSC)
Washington, DC 20207-0001
Phone: (301) 504-0990
E-mail: info@cpsc.gov
Web site: www.cpsc.gov
An independent Federal regulatory agency that works to save lives and keep people safe by reducing the risk of injuries and deaths associated with consumer products. CPSC helped to develop information about the placement and safe use of woodstoves.

Biodiesel

National Biodiesel Board
P.O. Box 104898
Jefferson City, MO 65110-4898
Phone: (800) 841-5849
Web site: www.biodiesel.org
An excellent source of current information on biodiesel.

Corn

Penn State University On-Line EnergySelector
Web site: http://energy.cas.psu.edu/energyselector
This neat on-line calculator shows you the fuel switching point between any two fuels.

Geothermal

Air-Conditioning & Refrigeration Institute (ARI)
4100 N. Fairfax Drive, Suite 200
Arlington, VA 22203
Phone: (703) 524-8800
E-mail: ari@ari.org
Web site: www.ari.org
A national trade organization representing air conditioner and air-source heat pump manufacturers. Their Web site is a good source of statistics, standards, and other information on air-source heat pumps and air conditioners.

Consortium for Energy Efficiency, Inc. (CEE)
One State Street
Suite 1400
Boston, MA 02109-3529
Phone: (617) 589-3949
E-mail: rfoster@cee1.org
Web site: www.cee1.org
A national nonprofit organization that promotes energy-efficient products. CEE has established recommended efficiency levels for heat pumps.

Geothermal Heat Pump Consortium, Inc.
701 Pennsylvania Avenue, N.W.
Third floor
Washington, DC 20004-2696
Phone: (202) 508-5500
E-mail: info@ghpc.org
Web site: www.ghpc.org
A nonprofit organization founded in 1994 to promote the growth of energy-efficient, environmentally friendly heating and cooling technology.

Geothermal Resources Council
P.O. Box 1350
Davis, CA 95617
Phone: (530) 758-2360
E-mail: grclib@geothermal.org
Web site: www.geothermal.org
A nonprofit organization dedicated to geothermal research, development, and education.

International Ground Source Heat Pump Association
499 Cordell South
Oklahoma State University
Stillwater, OK 74078-8018
Phone: (800) 626-4747, or (405) 744-5175
E-mail: mcarthl@okstate.edu (for membership application)
Web site: www.igshpa.okstate.edu
A nonprofit organization that offers a wide range of information on ground-source heat pumps. Their Web site also lists certified installers and has links to other sites for additional information.

Pellets

Combustion-Net
Web site: www.combustion-net.com
An excellent on-line, Canada-based clearinghouse of information on combustion technologies. Click on "Pellet Fuels" for information on wood pellets and other types of biomass pellets.

Pellet Fuels Institute
1601 North Kent Street, Suite 1001
Arlington, VA 22209
Phone: (703) 522-6778
E-mail: pfimail@pelletheat.org
Web site: www.pelletheat.org
A trade association that represents the fuel preparation and clean-burning technology of renewable biomass energy resources.

(See also listing for *HearthNet* and *Hearth, Patio & Barbecue Association* under "Wood")

Solar

American Solar Energy Society (ASES)
2400 Central Avenue, Suite G-1
Boulder, CO 80301
Phone: (303) 443-3130
E-mail: ases@ases.org
Web site: www.ases.org
This organization publishes *Solar Today* magazine, which often has articles on active solar heating systems.

Center for Renewable Energy and Sustainable Technologies (CREST)
1612 K Street, NW, Suite 202
Washington, DC 20006
Phone: (202) 293-2898
Web site: www.crest.org
A nonprofit organization (recently merged with the Renewable Energy Policy Project) dedicated to renewable energy, energy efficiency, and sustainable living.

Energy Efficiency and Renewable Energy Clearinghouse (EREC)
P.O. Box 3048
Merrifield, VA 22116
Phone: (800) 363-3732
Email: doe.erec@nciinc.com
Web site: www.eere.energy.com
The U.S. Department of Energy's comprehensive resource for energy efficiency and renewable energy information, plus access to more than 600 links and 80,000 documents.

Florida Solar Energy Center (FSEC)
1679 Clearlake Road
Cocoa, FL 32922-5703
Phone: (321) 638-1000
E-mail: webmaster@fsec.ucf.edu
Web site: www.fsec.ucf.edu
Provides highly regarded independent third-party testing and certification of solar hot-water systems and other solar or energy-efficient products.

Institute for Solar Living (ISL)
P.O. Box 836
Hopland, CA 95449
Phone: (707) 744-2017
Email: isl@rgisl.org
Web site: www.solarliving.org
A nonprofit educational organization that offers workshops on a wide variety of renewable-energy, energy-conservation, and building technologies.

National Aeronautics and Space Administration (NASA)
Science Enterprise Program
Web site: http://eosweb.larc.nasa.gov/sse/
Complete solar energy data for the entire planet, compliments of NASA. Click on "Meteorology and Solar Energy." After you log in for the first time, you can gather the information you need simply by pointing to your location on a world map.

Office of Energy Efficiency and Renewable Energy
U.S. Department of Energy
Web site: www.eren.doe.gov
An excellent source for renewable energy information. For a list of computer software for solar energy analysis and system design, try this link:
www.eren.doe.gov/consumerinfo/refbriefs/v101.html

Sustainable Buildings Industry Council (SBIC)
1331 H Street, N.W., Suite 1000
Washington, DC 20005
Phone: (202) 628-7400
E-mail: SBIC@SBICouncil.org
Web site: www.sbicouncil.org
Originally known as the Passive Solar Industries Council, the SBIC is a nonprofit organization whose mission is to advance the design, affordability, energy performance, and environmental soundness of buildings nationwide.

Wood

Chimney Safety Institute of America
2155 Commercial Drive
Plainfield, IN 46168
Phone: (317) 837-5362 or (800) 536-0118
E-mail: office@csia.org
Web site: www.csia.org/
A nonprofit educational foundation that has established the only nationally recognized certification program for chimney sweeps in the United States.

HearthNet
E-mail: webmaster@hearth.com
Web site: www.hearth.com
A comprehensive source of information on wood-, pellet-, coal-, and gas-burning hearth appliances and central heaters. This site offers links to manufacturers and local retailers, hundreds of articles, thousands of questions and answers, photographs, and much more.

Hearth, Patio & Barbecue Association
1601 North Kent Street, Suite 1001
Arlington, VA 22209
Phone: (703) 522-0086
Web site: www.hpba.org
An international trade association that represents the manufacturers, distributors, and retailers of hearth appliances, including (but not limited to) woodstoves, pellet stoves, fireplace inserts, and barbecues.

Masonry Heater Association of North America (MHA)
1252 Stock Farm Road
Randolph, VT 05060
Phone: (802) 728-5896
E-mail: bmarois@sover.net
Web site: www.mha-net.org
The MHA provides information on masonry heaters and will help you locate sources of masonry heaters and installers in the United States and Canada. This comprehensive Web site also offers photographs of masonry heater installations from coast to coast, a bookstore, library, news, member listings, links to other Web sites, and much more.

National Fire Protection Association
1 Batterymarch Park
P.O. Box 9101
Quincy, MA 02269-9101
Phone: (617) 770-3000
Web site: www.nfpa.org
This organization has developed standards for woodstove (and other heating devices) clearances from walls and ceilings that are used as the basis for many local building codes.

woodheat.org
Web site: www.woodheat.org
A noncommercial service in support of responsible home heating with wood. This site offers detailed information on wood-burning technologies, chimneys, firewood, safety, environmental issues, and much more.

GLOSSARY

absorber: the darkened surface of a solar storage element that absorbs solar heat.

active solar: employing solar collectors as well as pumps and controls that use electricity while gathering solar energy.

add-on: a heat pump or wood- or pellet-fired boiler or furnace that is attached to an existing fossil-fueled heating system; also called a dual unit.

air conditioner: a mechanical device used to cool a building. A heat pump can also serve as an air conditioner.

air-source heat pump: a type of heat pump that uses the natural heat of outdoor air as a heat source.

airtight: designed and built so that essentially all air entering a stove passes through one or more controllable air inlets. No stove is truly "airtight"; a more accurate term would be "air-controlled."

air-wash: a special stove-door design that keeps the glass viewing panel in the door clean.

ambient temperature: the prevailing or surrounding temperature.

analemma: a graduated scale in the shape of a figure eight that represents the motion of the sun during the year.

angle of incidence: the angle at which sunlight strikes a planar surface (often a solar collector or photovoltaic module).

annual fuel utilization efficiency (AFUE): a measure of heating-appliance efficiency that includes heating system start-up and cooling time and any other losses associated with its normal operation during the entire year.

antifreeze solution: a chemical-and-water mixture that prevents freezing.

aperture: the window or glazed area through which sunlight enters a building.

ash: the inorganic compounds remaining after combustion of a solid fuel. In the case of wood ash, the ash is comprised of trace minerals that were once part of trees.

auger: a screwlike mechanical device that moves pellet fuel from a storage bin to the firebox (burn pot) of a pellet-burning heating appliance.

back drafting: a potentially dangerous malfunction of a heating appliance that occurs when excessive negative pressure is generated in a house, causing the combustion gases to spill back into the living space.

backpuff: the result of the rapid burning of volatile gases within a firebox, evidenced by smoke and, sometimes, flames being forced back out of the stove.

backup: a secondary source of energy or heating that is used when the primary source is insufficient or not functioning.

baffle: a metal or ceramic partition designed to direct incoming air or exhaust gases through a specific path within a stove.

balance point: the outdoor temperature at which the heat loss of a house matches the heating output of a heat pump. Below the balance point, supplemental heating is required.

base burner: a type of woodstove that burns its fuel evenly over the entire base of the stove.

batch heater: a direct passive solar hot-water heating system that uses a black-painted water tank as both a collector and storage device.

bels: a logarithmic rating system for sound. One bel is equal to 10 decibels. Air conditioner and heat-pump operating sound is measured in bels.

biodiesel: a clean-burning diesel fuel made from natural, renewable sources such as new or used vegetable oil.

biomass: plant material, including wood, wood pellets, vegetation, grains, or agricultural waste, used as a fuel or energy source.

biosphere: the thin layer of water, soil, and air that supports all life on Earth.

boiler: an enclosed heat source (usually located in a basement, but sometimes outdoors) in which water is heated and then circulated for heating a house. Boilers can burn a wide variety of fuels.

bottom-feed: a design of burner in a pellet-fired heating appliance in which the pellets feed from underneath the fire.

Btu: British thermal unit(s); a quantitative measure of heat. The amount of heat required to raise the temperature of 1 pound of water by 1 degree Fahrenheit.

building envelope: the parts of a building—including walls, roof, insulation, windows, doors, and sometimes floors—that come in contact with the outside environment.

burn pot: the metal or ceramic device in the combustion chamber of a pellet-fired heating appliance in which fuel and air are mixed and primary combustion occurs.

bypass mode: the damper setting in a catalytic stove that bypasses the catalytic combustor during start-up or refueling.

bypass switch: an electrical switch, normally mounted near the main electrical circuit panel in a house, that disconnects the house circuitry from the main power lines to prevent a backup generator from accidentally sending electricity over the utility lines during a power failure.

carbon monoxide: a colorless, odorless, lethal gas that is the product of incomplete combustion of carbon.

catalytic combustor: a device employing a catalyst that can promote ignition of air/exhaust gas mixtures at a temperature that is lower than normally would be required.

catalytic woodstove: a woodstove that relies on a catalytic combustor to help ignite combustion gases.

central heating: a heating strategy that relies on a heat source (normally located in a basement) connected to a system of hot-water pipes or hot-air ducts that distribute heat throughout the living space of a house.

charcoaling: the process of forming charcoal from wood.

chimney flow reversal: a downdraft of air that blows down a chimney; usually the result of low fire temperature and gusty winds.

circulating pump: a mechanical device used to circulate water or a fluid through a system of heating pipes.

clean-burning technology: a woodstove design, first developed in New Zealand, that lowers emissions without the use of a catalytic converter.

cleanout tee: a T-shaped section of stovepipe that has a tight-fitting cover over one opening.

clinker: a fused mass of minerals (especially silica) that accumulates on the air holes of a burn pot in a pellet-fired heating appliance.

closed loop: a ground-source heat pump design that uses the heat in the ground as its heat source.

coefficient of performance (COP): a measure of a heat pump's efficiency.

coil: *See* heat exchanger.

combination system: a heating system that heats a house as well as domestic hot water. May also refer to a system that uses a combination of heat sources or distribution methods, such as a system that uses radiant floor heating as well as hot-water baseboards and/or forced hot air.

combustion chamber: the portion of a heating device where fuel is burned.

combustion efficiency: a measure of how completely a fuel has been burned.

compressor: a mechanical device at the heart of all heat pumps that causes the temperature of the refrigerant to rise. Types of compressors are reciprocating, rotary, and scroll.

concentrating collector: an active solar collector that can achieve high temperatures by using a mirrored surface to concentrate solar energy on an absorber (called a receiver).

conduction: the transfer of heat through a material without any appreciable movement of its molecules. For example, heat is transferred through a stove wall by conduction.

contraflow masonry heater/stove: a traditional Swedish and Finnish masonry heater/stove in which the hot combustion gases first flow up from the firebox to the top of the stove and then back down through several heat-exchange channels.

control: elements in a passive solar design that prevent summer overheating, such as roof and cantilever overhangs, awnings, blinds, or even landscaping and trellises. Other (more active) controls might include fan thermostats, vents, and other devices that assist or restrict heat flow.

convection: the movement of heat due to motion in liquids (such as water) and in gases (such as air) from warmer to cooler areas.

cooling cycle: the operational cycle of a heat pump when it is cooling a house.

cooling load calculation: the amount of cooling required to cool a house to a desired temperature; stated in Btu per hour or tons per hour.

cord: the basic measure for buying or selling firewood. A cord measures 8 feet long by 4 feet high by 4 feet deep and contains a nominal 128 cubic feet. *See also* face cord.

corn boiler/furnace/stove: a boiler, furnace, or stove that uses corn as fuel.

creosote: a deposit of condensed wood smoke, including vapors, tar, and soot, that condenses on stovepipes and chimney flues as a result of a smoldering fire.

damper: a movable valve or plate in a heating device that restricts or shuts off the draft.

defrost cycle: the operational cycle of an air-source heat pump when it is defrosting the coil of the outside unit. *See also* outside unit.

degree-day: a measure of the need for heating and cooling; a unit that represents one degree of declination from a given point (such as 65 degrees Fahrenheit) in the mean daily outdoor temperature.

design-day heat load: a quantity of heat (in Btu per hour) that is determined by adding together the heat-loss figures for each component of a house. This figure represents how much heat a heating system needs to provide. *See also* design-day heat loss.

design-day heat loss: the total amount of heat lost from a house on a "design day" for a particular geographic location (generally) assuming an indoor temperature of 72 degrees Fahrenheit.

desuperheater: a special heat exchanger in a ground-source heat pump; used to produce domestic hot water.

differential thermostat: a device that measures the difference in temperature between two or more locations in a heating system.

direct-expansion loop (DX system): a ground-source heat pump design in which a refrigerant passing through copper piping buried in the ground extracts heat from the soil.

direct gain: a passive solar heating strategy in which sunlight enters a house and strikes masonry walls and/or floors, which absorb and store the heat.

direct solar hot-water system: a solar heating system that directly heats water for domestic-hot-water or space-heating purposes.

discharge well: a well in a ground-source heat pump system; receives the "used" water from the heat pump.

distribution: Movement of heat from its source to the living space of a house. Also, the method by which solar heat circulates from the collection and storage points to different areas of a house.

draindown: a type of direct active solar hot-water heating system that relies on electronically controlled, motorized valves to cause water to drain from the collectors and exterior piping when outdoor temperatures approach freezing.

dual unit: a heat pump or wood- or pellet-fired boiler or furnace that is attached to an existing fossil-fueled heating system; also called an add-on.

efficiency: the measure of the ability of a heating system or heating device to utilize the full energy potential of a fuel; the ratio between the amount of usable heat produced and the amount of potential energy in the fuel.

emissions: all substances discharged into the air during combustion. *See also* gaseous emissions; particulate emissions.

energy-efficiency ratio (EER): a measure of a heat pump's cooling efficiency.

equipment cost: the initial cost of equipment, as for a heating system, including installation cost.

evacuated-tube collector: a solar collector used to heat water for home heating applications that require higher temperatures than standard collectors can provide.

expansion device: the heat pump component that releases the pressure created by the compressor, causing the temperature of the refrigerant to drop.

expansion tank: a tank that compensates for the varying pressure caused by the heating and cooling of water in a hot-water heating system.

face cord: a measure of firewood that is 4 feet high by 8 feet long and as deep as the length of a single log in the cord. *See also* cord.

factory-built fireplace: a manufactured metal fireplace (also known as a zero-clearance fireplace) using a metal chimney that can be easily installed in almost any location in a home.

fines: very fine dust (considered a negative attribute) that accompanies some pellet fuels and grains.

fireback: the rear portion of the firebox in a fireplace.

firebox: the area within a heating appliance used to contain the burning fuel.

firebrick: a brick made of materials capable of withstanding high temperatures for extended periods of time; used to build fireplaces and to line furnaces and heating stoves.

fireplace: a generally rectangular opening in the lower part of a chimney that can hold an open fire.

fireplace insert: a solid-fuel combustion device that is placed in a fireplace cavity. A fireplace insert offers greatly improved efficiency over an open fireplace by reducing excess air through an air-control device.

flat-plate collector: an insulated metal box with a dark-colored absorber plate and a glass or plastic cover; used to heat liquids or air in an active or passive solar heating system.

forced hot-air system: a heating system that consists of a heat source and ductwork to distribute heated air through a house with the assistance of a blower.

fossil fuel: an organic, energy-rich substance formed from the long-buried remains of prehistoric life. Fossil fuels are considered nonrenewable, and their use contributes to air pollution and global warming.

four-way valve: a valve that controls the direction of flow within a heat pump depending on whether the heat pump is heating or cooling a house; also called a magic valve.

front-to-back burner: a type of woodstove that burns firewood from one end of the log to the other, like a cigar burns.

fuel cell: a conversion device that uses fuel to create electricity.

fuel-limited: describes a heating device such as an open fireplace or masonry stove that uses the amount of fuel burned as the main control on heat output.

furnace: an enclosed heat source (usually located in a basement) in which heat is produced to heat a house, usually via hot air. Furnaces can burn a wide variety of fuels.

gaseous emissions: gases discharged into the air during combustion, typically including carbon dioxide, carbon monoxide, water vapor, and hydrocarbons.

gasification: the process of converting a solid fuel such as wood or pellets into a combustible gas.

gasohol: a fuel blend of ethanol made from fermented corn and gasoline.

geothermal: having to do with heat in the air, ground, or ground water. Used in the context of heating systems, geothermal refers to systems that pump heat either from or into the ground or outside air in order to heat or cool a building.

grate: a frame, bed, or platform within a stove or furnace on which a fire is kindled.

gravity drainback: a design that relies on gravity to cause water to drain out of the collectors and exterior piping of a direct active solar hot-water heating system when outdoor temperatures approach freezing.

greenhouse effect: the heating of the atmosphere that results from the absorption of re-radiated solar radiation by certain gases, especially carbon dioxide and methane.

green wood: freshly cut firewood with a high moisture content. When expressed as a percentage of the weight of the wood when dry, the moisture content of green wood can be 100 percent. *See also* seasoned wood.

ground-source heat pump: a type of heat pump that uses the natural heat of the ground, ground water, or surface water as a source of heat. Also called GeoExchange or Earth-energy systems.

hearthpad: a special insulated pad placed under a woodstove to protect the combustible flooring beneath from catching fire.

heat emitter: a device used to transfer heat from a distribution system to the living space in a house.

heat exchange: the process whereby heat moves from hot areas or objects to cooler areas or objects; also called heat transfer. *See also* conduction, convection, radiation.

heat exchanger: any device designed to transfer heat. An area of (or within) a stove, furnace, or fireplace insert where heat in hot exhaust gases can be transferred to room air or to incoming combustion air. In a solar hot-water heater, the heat exchanger usually transfers the heat from an antifreeze fluid to water. In a heat pump, the heat exchanger (also called a coil) performs the same heat-transfer function between air, refrigerant, or water, depending on the type of system.

heatilator: a generic term for factory-built steel fireplace units that include hot-air ducts in their design. Also, the name of a manufacturer of factory-built fireplace units.

heating cycle: the operational cycle of a heat pump when it is heating a house.

heating-season performance factor (HSPF): a measure of the total heat output (expressed in Btu) of a heat pump over an entire season.

heating value: the amount of heat that has been generated per unit of fuel when the fuel has completely combusted.

heat pump: a mechanical device, with a compressor as its main component, that heats or cools a house using one of several heat sources including the ground, a body of water, or the ambient air.

heat shield: a noncombustible protector used around appliances, smoke pipe, or chimneys.

heat sink: A component—such as masonry, concrete, stone, or water—inside a structure that absorbs heat and then radiates heat slowly when the surrounding air falls below its temperature.

heat transfer: the process whereby heat moves from hot areas or objects to cooler areas or objects; also called heat exchange. *See also* conduction, convection, radiation.

horizontal loop: a type of closed-loop design in a ground-source heat pump in which the heat-transfer piping is installed horizontally in trenches dug in the ground.

HVAC: abbreviation for heating, ventilation, and air-conditioning.

hybrid system: any home heating system that combines two or more different heating functions or strategies.

hydronic: a system of heating or cooling that involves the transfer of heat by a circulating fluid in a closed system of pipes.

hydronic baseboard heater: a baseboard unit mounted along an exterior wall, usually under windows, through which heat is distributed by hot water supplied from a boiler or water heater.

hydronic floor: a floor mass, typically concrete, with embedded tubes through which heated fluid is pumped. The flooring absorbs the heat from the fluid and gradually radiates heat upward into the room.

ignition system: an automatic ignition device in some pellet-fired heating appliances that starts the fire automatically or at the touch of a button.

ignition temperature: the temperature at which a substance will spontaneously ignite in the presence of sufficient oxygen.

incident solar radiation: the amount of sunlight falling on a place, usually an average representing unclouded sunshine per day averaged over a year. Also called solar insolation.

indirect gain: a passive solar heating strategy that captures the sun's heat in a thermal mass, usually a Trombe wall, located between the sun and the living space of the house. *See also* Trombe wall.

indirect solar hot-water system: a solar heating system that indirectly heats water for domestic-hot-water or space-heating purposes through the use of a heat exchanger and antifreeze.

infiltration: the leakage of cold air through cracks, crevices, or other openings in the structure of a house.

infrared: refers to light just outside the visible spectrum, normally associated with heat radiation.

inside unit: the part of an air-source heat pump that is located inside of a house; contains the inside heat exchanger and fan.

insulation: any material used to keep energy, especially heat, from escaping or conducting. In buildings, insulation keeps heat in during the winter and out during the summer.

isolated gain: a passive solar heating strategy that generally refers to trapping solar heat in a sunspace (solar room or solarium), which can be included in the original design of a house or added on later.

joule: the basic unit of heat energy in the metric system. One joule equals 0.00095 Btu.

kilocalorie: a quantitative measure of heat equal to 1,000 calories, where 1 calorie is equal to the amount of heat needed to raise the temperature of 1 gram of water by 1 degree Celsius. One kilocalorie equals 3,968 Btu.

kilowatt: one thousand watts; ten 100-watt lightbulbs consume one kilowatt of electricity. One kilowatt equals 3,415 Btu. *See also* watt.

life-cycle cost: the estimated cost of owning and operating a (heating) system over its useful life.

limb wood: small diameter firewood that comes from the limbs, rather than the trunks, of trees; generally preferred for use in masonry heaters.

log set: an artificial log that simulates a burning log; used in some pellet-fired heating appliances.

low-e (low-emissivity) windows: a form of high-tech glass incorporating a coating that reduces the passage of radiant energy.

low-voltage: in electrical terms, less than house current, typically 12 or 24 volts.

magic valve: *See* four-way valve.

masonry heater: a large, heavy heater, usually constructed out of masonry, that absorbs the heat from a quick, hot fire and then radiates the heat back into a house for many hours after the fire has burned out. Also called a Kachelöfen or tile stove or a Russian, Siberian, Finnish, or Swedish stove or fireplace.

MBH heating capacity: the rated capacity of heating equipment, measured in thousands of Btu per hour.

mean radiant temperature: the average temperature of the surfaces (such as walls, floor, ceiling, and windows) in a room.

noncatalytic woodstove: a stove that includes a firebox, air controls, and baffles to recirculate smoke for more efficient combustion. Also called a clean-burning or recirculating woodstove.

open-discharge method: a ground-source heat pump design that discharges the "used" water from the system into a drain tile, stream, river, pond, or lake.

open system: a ground-source heat pump design that uses the heat in a body of water—usually a well, but sometimes a pond or stream—as its heat source.

operating cost: the cost of operating a system or appliance over its useful life.

orientation: placement with respect to the cardinal points (north, south, east, and west).

outside-air kit: a device that provides additional combustion air for a variety of heating appliances.

outside unit: the part of an air-source heat pump that contains the outside heat exchanger, the fan, and usually the compressor; this unit is located outside of a house.

overall efficiency: the ratio between the heat that enters a room as the result of burning a fuel and the total heat available from that fuel. Overall efficiency has two components, combustion efficiency and heat-transfer efficiency. *See also* combustion efficiency.

panel: any flat modular structure; solar panels collect solar energy by a number of different strategies.

particulate emissions: minute particles discharged into the air during combustion, typically fly ash, tar aerosols, carbonaceous soot, and various organic molecules.

passively heated: heated by the sun.

passive solar: a strategy for design and construction of a house that maintains a comfortable inside temperature by admitting, storing, and preserving the heat from sunlight.

payback: the amount of time required for the total of the equipment and operating costs of a more efficient unit to fall below the total cost of a conventional unit; important when comparing two systems, usually a conventional, low-priced, high-maintenance-cost system and a higher-priced, cheaper-to-operate system. *See also* equipment cost; operating cost.

peak sun hour: the unit of measure of solar insolation. *See also* solar insolation.

pellet boiler/furnace/stove: a boiler, furnace, or stove that uses biomass pellets (or sometimes corn) as its source of fuel.

pellets: biomass fuels generally made from sawdust and ground wood chips. Ground wood chips are waste materials from trees that have been harvested to make furniture, lumber, and other products.

phase-change liquid: a liquid used as a heat-transfer fluid in heat pumps and sometimes in active solar heating systems instead of water or antifreeze solution.

photosynthesis: a process by which plants and other organisms use light to convert carbon dioxide and water into a simple sugar. Photosynthesis provides the basic energy source for almost all organisms.

photovoltaics (PVs): modules that utilize the photovoltaic effect to generate electricity.

plenum: the largest element in a hot-air duct system, usually a large rectangular box made of sheet metal.

primary air: the air entering a stove, used to maintain combustion in the firebox.

product of combustion: a substance formed during combustion of a fuel. The products of complete combustion are carbon dioxide and water. Products of incomplete combustion can include carbon monoxide, hydrocarbons, soot, tars, and other substances.

radiant heating: a heating system that works by heating surfaces (usually floors, panels, or ceilings), from which heat eventually radiates into the living space.

radiant stove: a stove that emits most of its heat in the form of infrared radiation.

radiation: the transfer of heat in the form of rays or waves. Also, the radiation of heat and light from a burning object or fuel.

recirculating: a design strategy in which hot water from the storage tank of a direct active solar hot-water heating system is pumped through the collectors and exterior piping when outdoor temperatures approach freezing.

refractory: resistant to heat; referring to a ceramic material used to line stoves and furnaces to protect metal parts from extreme heat and oxidation.

relative humidity: the amount of water vapor in air at a particular temperature, expressed as a percentage of the maximum amount of water that the air could hold at that temperature.

renewable energy: an energy source that renews itself; the sun is the best example.

retrofit: install in (or on) a house new equipment that was not originally designed for the house.

R-value: the measure of a material's resistance to thermal transfer.

SDHW system: a solar domestic-hot-water system; a system for heating household (domestic) hot water using solar energy.

seasonal energy-efficiency ratio (SEER): the measure of seasonal cooling efficiency of a heat pump.

seasoned wood: firewood that has been dried over a period of time to reach a moisture content of between 15 and 25 percent. *See also* green wood.

seasoning: the initial break-in procedure for a new stove that gives the stove metal a chance to expand and contract gradually and that introduces the stove to repeated heating.

secondary combustion: combustion of unburned gases emitted from the primary mass of a fire; this combustion usually occurs in a secondary chamber.

set-back thermostat: a device that combines a clock and a thermostat for automatic heat control.

shield: a specially shaped metal component that seals off the space between a fireplace insert and the mouth of the fireplace; also called a shroud.

slip-connector: a special section of stovepipe that telescopes, making stove installation and pipe-cleaning operations easier.

solar aperture: the opening to the south of a site (in the Northern Hemisphere) across which the sun passes; trees, buildings, and mountains may obstruct the solar aperture, which changes with the seasons.

solar assist: using the sun's energy to help a process proceed.

solar collector: a device used to collect solar energy.

solar collector array: a group of solar collectors.

solar energy: energy from the sun; the most environmentally friendly energy source available.

solar fraction: the portion of heat that can be provided by a typical active solar home heating system.

solar gain: the absorption of heat from the sun.

solar hot-water heating: direct or indirect use of heat from the sun to heat hot water for domestic-hot-water and space-heating purposes.

solar insolation: the amount of sunlight falling on a place, usually an average representing unclouded sunshine per day averaged over a year; also called incident solar radiation.

solar panel: any flat surface designed to gather solar energy.

solar room heater: a simple solar hot-air heating system (sometimes called a wall heater) that is only intended to heat one room. These systems can be either active or passive.

southern exposure: the side of a house that most directly faces south.

stack loss: heat and potential heat that is lost up a smokestack or chimney.

standing column well: a type of deep-drilled well that is part of an open-system heat pump design; water for the heat pump is drawn from the bottom of the well while the return water is discharged into the top of the well.

starter gel: a product used to start the fire in some models of pellet- and corn-fired heating appliances.

steady-state efficiency: the ratio of the heat actually available for use in the distribution system of a heating appliance to the amount of heat potentially available in the fuel.

stove: a heating device that burns fuel to produce heat.

sun-tempered house: a house that receives some of its space heating from the sun.

supplementary heating: a heating system or strategy that provide additional or supplementary heat for a primary heating system.

sustainable: referring to material or energy sources that, if carefully managed, will provide at current levels indefinitely.

temperature stratification: the difference between the air temperature at floor level and ceiling level in a room.

terawatt: one trillion watts. *See also* watt.

thermal conductivity: the measure of a material's ability to conduct heat. Metal has a high thermal conductivity and is a good medium for transferring heat.

thermal mass: solid, usually masonry (but sometimes water-filled), components inside a house that absorb heat and then radiate heat slowly when the temperature of the surrounding air falls below their temperature.

thermosiphoning system: a plumbing system that takes advantage of the fact that cold water is heavier than hot water and that will cause water to circulate as long as there is a heat differential.

thermostat: a device that automatically responds to temperature changes in a home-heating system or air conditioner and that controls heating or cooling output.

thimble: a device through which stovepipe passes into a chimney.

ton: a measure of heat pump size; a 1-ton heat pump can generate 12,000 Btu of cooling per hour at an outdoor temperature of 95 degrees Fahrenheit or 12,000 Btu of heat output per hour at 47 degrees Fahrenheit.

top damper: a damper operated by a long stainless steel cable that runs up the flue of a chimney. A top damper completely seals the flue from the outside environment when the fire in a fireplace has completely burned out.

top feed: a design of burner in a pellet-fired heating appliance in which the pellets fall into the fire from above.

Trombe wall. An 8- to 16-inch-thick masonry wall constructed in a room on the south side of a passive solar house and located 1 inch or less away from the exterior glass. The wall heats up and gradually releases heat into the room over a period of hours.

type "L" or "PL" vent pipe: a special double-wall design of pipe used to vent pellet-fired heating appliances.

unitary space heater: a heating device intended to only heat the room or area in which it is located.

vertical loop: a type of closed-loop design in a ground-source heat pump in which the heat transfer piping is installed vertically in bored holes in the ground.

volatiles: gaseous and liquid materials that are driven from wood as they are heated to temperatures between 500 degrees and 600 degrees Fahrenheit. Wood volatiles usually contain various hydrocarbons and can represent up to one-third or more of the wood's energy.

wall heater: a passive solar heating design that uses part of a south-facing wall as a solar collector.

wall thimble: a device installed in combustible walls, through which stovepipe passes; intended to help protect the walls from catching fire due to stovepipe heat.

water jacket: a device in a boiler that transfers heat from the fire to the hydronic heat distribution system or a similar device that can be installed in wood-fired kitchen ranges to heat water for domestic purposes.

watt: a measure of electrical power, calculated as the mathematical product of voltage times amperage.

wood gasification: a process where wood is heated to drive off volatile gases, which are then burned in a secondary combustion chamber at temperatures up to 2,000 degrees Fahrenheit.

woodstove: a stove that uses wood as the source of fuel.

BIBLIOGRAPHY

Biodiesel

Tickell, Joshua. *From the Fryer to the Fuel Tank: The Complete Guide to Using Vegetable Oil as an Alternative Fuel.* Sarasota, Florida: GreenTeach Publishing, 2000.

Geothermal

Langley, Billy C. *Heat Pump Technology.* Upper Saddle River, New Jersey: Prentice Hall, 2002.

Solar

Chiras, Dan. *The Solar House: Passive Heating and Cooling.* White River Junction, Vermont: Chelsea Green Publishing Co., 2002.

Crosbie, Michael J., ed. *The Passive Solar Design and Construction Handbook.* New York: John Wiley & Sons, 1997.

Dresser, Peter van. *Passive Solar House Basics.* Santa Fe, New Mexico: Ancient City Press, 1996.

Freeman, Mark. *The Solar Home: How to Design and Build a House You Heat with the Sun.* Mechanicsburg, Pennsylvania: Stackpole Books, 1994.

Harland, Edward. *Eco-Renovation: The Ecological Home Improvement Guide.* White River Junction, Vermont: Chelsea Green Publishing Co., 1999.

Kachadorian, James. *The Passive Solar House.* White River Junction, Vermont: Chelsea Green Publishing Co., 1997.

Potts, Michael. *The New Independent Home: People and Houses that Harvest the Sun, Wind, and Water.* White River Junction, Vermont: Chelsea Green Publishing Co., 1999.

Roy, Rob. *The Complete Book of Underground Houses: How to Build a Low-Cost Home.* New York: Sterling Publishing Co., 1994.

Sklar, Scott and Sheinkopf, Kenneth (contributor). *Consumer Guide to Solar Energy: New Ways to Lower Utility Costs, Cut Taxes, and Take Control of Your Energy Needs.* Chicago: Bonus Books, 2002.

Masonry Stoves

Barden, Albert. *Finnish Fireplace Construction Manual (with Updates).* Norridgewock, Maine: Maine Wood Heat Company, Inc., 1988.

Barden, Albert and Hyytiäinen, Heikki. *Finnish Fireplaces: Heart of the Home.* Norridgewock, Maine: Maine Wood Heat Company, Inc., 1988.

Lyle, David. *The Book of Masonry Stoves: Rediscovering an Old Way of Warming.* White River Junction, Vermont: Chelsea Green Publishing Co., 1997.

Wood

Bushway, Stephen, and Twitchell, Mary (Editor). *The New Woodburner's Handbook: A Guide to Safe, Healthy & Efficient Woodburning.* Pownal, Vermont: Storey Books, 1992.

Johnson, Dave. *The Good Woodcutter's Guide: Chain Saws, Portable Sawmills, and Woodlots.* White River Junction, Vermont: Chelsea Green Publishing Co., 1998.

Ritchie, Ralph W. *All That's Practical about Wood: Stoves—as a Fuel—Heating.* Springfield, Oregon: Ritchie Unlimited Publications, 1998.

Thomas, Dirk. *The Woodburner's Companion: Practical Ways of Heating with Wood.* Chambersburg, Pennsylvania: Alan C. Hood & Co., 2000.

INDEX

I

J

K

L

M

N